D1256658

SELECTED TOPICS IN GRAPH THEORY 3

SELECTED TOPICS IN GRAPH THEORY 3

Edited by

LOWELL W. BEINEKE

Department of Mathematical Sciences
Indiana University—Purdue University at Fort Wayne
Indiana, U.S.A.

and

ROBIN J. WILSON

Faculty of Mathematics
The Open University, England

1988

ACADEMIC PRESS LIMITED
Harcourt Brace Jovanovich, Publishers
London San Diego New York Boston
Sydney Tokyo Toronto

ACADEMIC PRESS LIMITED
24/28 Oval Road,
London NW1 7DX

United States Edition published by
ACADEMIC PRESS INC.
San Diego
CA 92101

British Library Cataloguing in Publication Data

Selected topics in graph theory.
Vol. 3
1. Graph theory
I. Beineke, Lowell W. II. Wilson, Robin J.
511′.5 QA166

ISBN 0-12-086203-4

Typeset by Setrite Typesetters Ltd., Hong Kong
Printed in Great Britain at the University Press, Cambridge

Preface

One of the consequences of the recent rapid expansion in graph theory is that it has become increasingly difficult to ascertain what is currently known about a particular topic in the field. In view of this, we felt that it would be worthwhile to collect together a series of expository surveys written by a distinguished group of authors and covering various areas of graph theory, in the hope that such a collection might prove useful to professional graph theorists, to newcomers to the field, and to experts in other fields who may want to learn about specific topics. The result of this was *Selected Topics in Graph Theory* which appeared in 1978; we refer to this book as **ST1**.

Because that book was well received, and because it included only a limited number of topics, it was felt that it would be worthwhile to produce two further volumes. In these books, the selection of topics chosen is entirely our own, and we are well aware that many important areas have still had to be omitted. Our choices were made on the basis of several criteria, including the need for surveys on particular topics, the timeliness of certain areas, suggestions from colleagues and friends, and, of course, our own preferences. In these books, the emphasis is on 'pure' graph theory, and the reader who is primarily interested in applications of graph theory should refer to our companion volume *Applications of Graph Theory*. It is a collection of original expository writings covering applications of graph theory to a wide variety of subjects, ranging from communications networks and chemistry to geography and architecture.

In these books we have attempted, as far as possible, to impose a uniform terminology and notation throughout in the belief that this will aid the reader in going from one chapter to another. It should also make the books more accessible to groups using them for advanced courses or seminars. In order to give the chapters a fairly consistent style, we asked our contributors to undergo the ordeal of having early drafts of their chapters subjected to severe criticism. We believe that this resulted in a considerable improvement in the final versions, and we should like to express our thanks and appreciation to all our contributors for their co-operation in this, and in particular for their tolerance and their willingness to put up with our idiosyncracies.

We should also like to thank the reviewers of the early drafts of chapters for their helpful and pertinent comments, the mathematics departments of the Open University, Oxford University, Colorado College, and Indiana University–Purdue University at Fort Wayne, for their support, Academic Press for its encouragement and co-operation, and (finally) our wives and children, who have had to put up with us during the writing of this book.

L.W.B.
R.J.W.

Notes on Contributors

Lowell Beineke is Professor of Mathematics at Indiana University–Purdue University at Fort Wayne, where he moved after receiving his Ph.D. from the University of Michigan. He has contributed to a wide variety of areas in graph theory, including topological graph theory, line graphs, tournaments and digraphs, and he is on the Editorial Board of the *Journal of Graph Theory*. He is co-author/editor (with R. J. Wilson) of *Selected Topics in Graph Theory 1, 2* (Academic Press, 1978, 1983) and *Applications of Graph Theory* (Academic Press, 1979). *Present address: Department of Mathematical Sciences, Indiana University–Purdue University, Fort Wayne, Indiana 46805, U.S.A.*

Claude Berge is Professor of Mathematics at the University of Paris 6, and Director of Research at the C.N.R.S. (National Centre for Scientific Research). His main research interests lie in topology, combinatorics and graph theory, and he has written many papers and books in these areas, including *Theory of Graphs and its Applications* (Methuen and Wiley, 1961), *Principles of Combinatorics* (Academic Press, 1971), *Graphs and Hypergraphs* (North-Holland, 1973) and *Graphes* (Gauthier Villars, 1983), and is co-author (with A. Ghouila-Houri) of *Programming, Games and Transportation Networks* (Methuen, 1965). *Present address: 10 rue Galvani, 75017 Paris, France.*

Fan R. K. Chung is Division Manager of Mathematics, Information Sciences and Operations Research at Bell Communications Research. After receiving her Ph.D. from the University of Pennsylvania, she worked at Bell Laboratories at Murray Hill before she moved to Bellcore to build the discrete mathematics research group there. She has worked in a wide variety of areas in graph theory, combinatorics and algorithmic analysis, with particular interest in Ramsey numbers, graph decompositions, graph embeddings, universal and unavoidable graphs, diameters, combinatorial geometry and network optimization. She is now the Managing Editor of the *Journal of Graph Theory*. *Present address: Room 2L-387, Bell Communications Research, 435 South Street, Morristown, New Jersey 07960, U.S.A.*

Ronald Graham is Director of the Mathematical Sciences Research Center at AT&T Bell Laboratories and University Professor of Mathematical Sciences at Rutgers University. He is a member of the National Academy of Sciences and is a Fellow of the American Academy of Arts and Sciences. His professional interests include combinatorics, graph theory, number theory, geometry and various areas of theoretical computer science. He is the author of *Rudiments of Ramsey Theory* (Amer. Math. Soc., 1981), co-author (with B. L. Rothschild and J. H. Spencer) of *Ramsey Theory* (Wiley, 1980) and co-author (with P. Erdős) of *Old and New Problems and Results in Combinatorial Number Theory* (Univ. Geneva, 1980). *Present address: Mathematical Sciences Research Center, AT&T Bell Laboratories, 600 Mountain Avenue, Murray Hill, New Jersey 07974, U.S.A.*

François Jaeger holds a research position at the Centre National de la Recherche Scientifique and has worked in Grenoble since 1970. He received his State Doctorate in 1976. His main research interests are in graph theory and matroid theory (with a special interest in the interplay between these topics), linear algebra and geometry. *Present address: Laboratoire de Structures Discrètes, IMAG, BP68, 38402—St Martin d'Heres Cedex, France.*

Joseph Malkevitch is Professor of Mathematics at York College, CUNY, New York. He received his Ph.D. at the University of Wisconsin in 1969, and his research interests lie in geometry, graph theory, combinatorics and mathematics education. He has written monographs on planar graphs and the mathematical theory of elections, and is co-author (with W. Meyer) of *Graphs, Models, and Finite Mathematics* (Prentice-Hall, 1974) and a co-author for *For all Practical Purposes* (W.H. Freeman, 1988) to accompany the associated PBS television series. *Present address: Department of Mathematics, York College, CUNY, Jamaica, New York 11451, U.S.A.*

Ronald Read is Professor in the Department of Combinatorics and Optimization at the University of Waterloo. He has contributed to several areas of graph theory, especially in those aspects relating to algorithms and computing, and is also interested in enumeration problems in organic chemistry. He is the editor of *Graph Theory and Computing* (Academic Press, 1972). *Present address: Department of Combinatorics and Optimization, The University of Waterloo, Waterloo, Ontario N2L 3G1, Canada.*

Carsten Thomassen is Professor of Discrete Mathematics at the Technical University of Denmark. He received his Ph.D. from the University of Waterloo, after which he was appointed lektor at Aarhus University for a while. He is on the Editorial Boards of the *Journal of Graph Theory*, the *Journal of Combinatorial Theory (B)*, *Discrete Math.*, *Combinatorica*, and *Aequationes Mathematicae*. His diverse combinatorial interests include infinite graphs, tournaments, planarity, connectivity and extremal graphs. *Present address: Mathematics Institute, Bygn. 303, The Technical University of Denmark, 2800 Lyngby, Denmark.*

William T. Tutte studied graph theory while a chemistry student at Trinity College, Cambridge. He taught mathematics at the University of Toronto from 1948 to 1962 and at the University of Waterloo from 1962 to 1985, specializing in combinatorics at Waterloo. His many papers cover the entire range of pure graph theory and include distinguished contributions to matroid theory, enumeration, factorization and connectivity. He is the author of *Connectivity in Graphs* (Toronto, 1966), *Introduction to the Theory of Matroids* (American Elsevier, 1971) and *Graph Theory* (Cambridge University Press, 1985). *Present address: 16 Bridge Street, W. Montrose, Ontario N0B 2V0, Canada.*

Dominic Welsh is Fellow and Tutor in Mathematics at Merton College, Oxford University. His main researches have been in probability theory, matroid theory, percolation theory and computational complexity. He is the author of *Matroid Theory* (Academic Press, 1976), co-author (with G. R. Grimmett) of *Probability, an Introduction* (Oxford University Press, 1986), editor of *Combinatorial Mathematics and its Applications* (Academic Press, 1971), and co-editor (with D.R. Woodall) of *Combinatorics* (Inst. Math. Appl., 1972). *Present address: Merton College, Oxford OX1 4JD, England.*

Robin Wilson is Senior Lecturer in Mathematics at the Open University. He received his Ph.D. in number theory from the University of Pennsylvania, and shortly afterwards became interested in graph theory and combinatorics. His main areas of interest are in edge-colorings of graphs, algebraic graph theory and the history of graph theory and combinatorics. He is the author of *Introduction to Graph Theory* (Longman/Academic Press, 1972, 1985), co-author (with N. L. Biggs and E. K. Lloyd) of *Graph Theory 1736–1936* (Oxford University Press, 1976), co-author (with Stanley Fiorini) of *Edge-Colourings of Graphs* (Pitman, 1977), and has edited or co-edited six books, including (with L. W. Beineke) *Selected Topics in Graph Theory 1, 2* (Academic Press, 1978, 1983) and *Applications of Graph Theory* (Academic Press, 1979). *Present address: Faculty of Mathematics, The Open University, Milton Keynes MK7 6AA, England.*

Contents

1 Introduction

In this introductory chapter we present those definitions and theorems in graph theory which will be assumed throughout the rest of the book. Further explanations of these terms, together with the proofs of stated results, will be found in the standard texts in the subject (see, for example, [**1**], [**2**], [**5**] and [**9**]), although not all of the terminology is completely standardized. Definitions and results not included here are introduced later on as they are needed, or may be found in the above-mentioned texts.

Graphs

A **graph** G is a pair $(V(G), E(G))$, where $V(G)$ is a finite non-empty set of elements called **vertices**, and $E(G)$ is a finite set of distinct unordered pairs of distinct elements of $V(G)$ called **edges** (see Fig. 1). We call $V(G)$

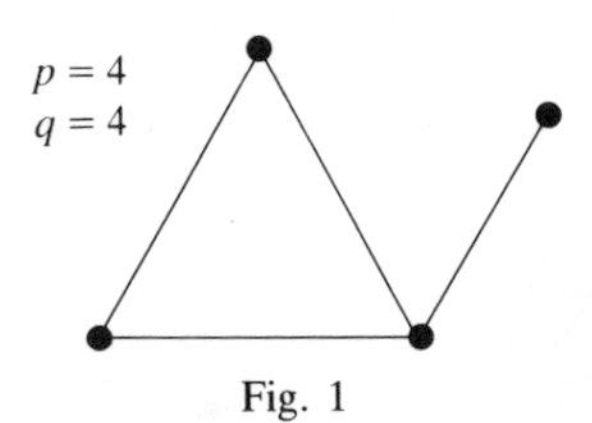

Fig. 1

the **vertex-set** of G, and $E(G)$ the **edge-set** of G; when there is no possibility of confusion, these are sometimes abbreviated to V and E, respectively. The number of vertices of G is called the **order** of G, and will usually be denoted by p; the number of edges of G will generally be denoted by q. For convenience, we usually denote the edge $\{v, w\}$ (where v and w are vertices of G) by vw.

If $e = vw$ is an edge of G, then e is said to **join** the vertices v and w, and these vertices are then said to be **adjacent**. In this case, we also say that e is **incident** to v and w, and that w is a **neighbor** of v; the **neighborhood**

GRAPH THEORY, 3
ISBN 0-12-086203-4

of v, denoted by $N(v)$, is the set of all vertices of G adjacent to v. Two edges of G incident to the same vertex are called **adjacent edges**. An **independent set of vertices** in G is a set of vertices of G, no two of which are adjacent, and the size of the largest such set is called the **independence number** of G. Similarly, an **independent set of edges**, or **matching**, in G is a set of edges of G no two of which are adjacent, and the size of the largest such set is called the **edge-independence number** of G. An independent set of edges which includes every vertex of G is called a **1-factor**, or **complete matching**, in G.

Two graphs G and H are said to be **isomorphic** (written $G \cong H$) if there is a one-to-one correspondence between their vertex-sets which preserves the adjacency of vertices. An **automorphism** of G is a one-to-one mapping φ of $V(G)$ onto itself with the property that $\varphi(v)$ and $\varphi(w)$ are adjacent if and only if v and w are. The automorphisms of G form a group $\Gamma(G)$ under composition, called the **automorphism group** of G; $\Gamma(G)$ is said to be **transitive** if it contains transformations mapping each vertex of G to every other vertex.

Some Variations

If, in the definition of a graph, we remove the restriction that the edges must be distinct, then the resulting object is called a **multigraph** (see Fig. 2); two or more edges joining the same pair of vertices are then called **multiple edges**. If M is a multigraph, its **underlying graph** is the graph obtained by replacing each set of multiple edges by a single edge; for example, the underlying graph of the multigraph in Fig. 2 is the graph in Fig. 1.

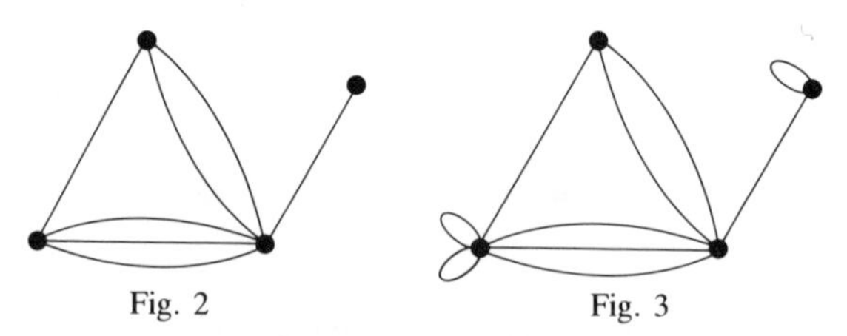

Fig. 2 Fig. 3

If we also remove the restriction that the edges must join distinct vertices, and allow the existence of **loops**, then the resulting object is called a **general graph**, or **pseudograph** (see Fig. 3). When concentrating our attention on graphs, as opposed to multigraphs or general graphs, we shall sometimes use the term "simple graph" to emphasize the fact that we are excluding loops and multiple edges.

A graph in which one vertex is distinguished from the rest is called a **rooted graph**. The distinguished vertex is called the **root-vertex**, or simply the **root**, and is often indicated by a small square (see Fig. 4). A **labeled graph** of order p is a graph whose vertices have been assigned the numbers 1, 2, ..., p in such a way that no two vertices are assigned the same number (see Fig. 5).

We can also consider directed graphs, in which the word "unordered" in the definition of a graph is replaced by "ordered". More formally, we define a **digraph** D to be a pair $(V(D), A(D))$, where $V(D)$ is a finite non-empty set of elements called **vertices**, and $A(D)$ is a finite set of distinct ordered pairs of distinct elements of $V(D)$ called **arcs** (see Fig. 6);

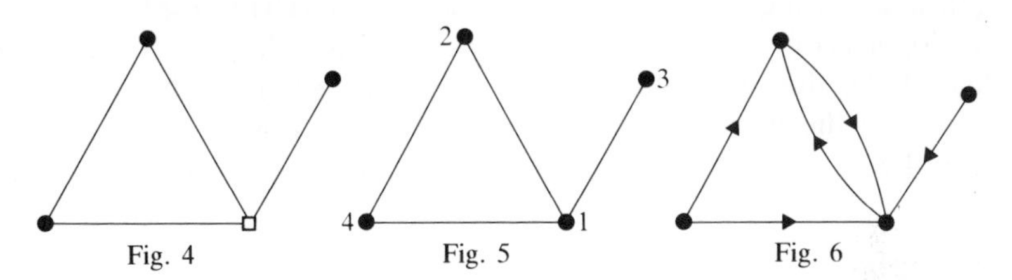

Fig. 4 Fig. 5 Fig. 6

we shall usually denote the arc (v, w) (where v and w are vertices of D) by vw. If $e = vw$ is an arc of D, then we say that v and w are **adjacent**, and that e is **incident from** v and **incident to** w. A pair of arcs of the form vw and wv is called a **symmetric pair**. If D is a digraph, the **underlying graph** of D is the graph or multigraph obtained from D by replacing each arc by an (undirected) edge joining the same pair of vertices. A **complete symmetric digraph** is a digraph in which every two vertices are joined by exactly two arcs, one in each direction, and a **tournament** is a digraph in which every two vertices are joined by exactly one arc; tournaments are discussed at length in **ST1**, Chapter 7.

We also define a **hypergraph** G to be a pair $(V(G), E(G))$, where $V(G)$ is a finite non-empty set of vertices, and $E(G)$ is a finite set of unordered sets of distinct elements of $V(G)$ called **edges**. If all edges in $E(G)$ have the same cardinality k, then G is called a **k-uniform hypergraph**. Hypergraphs are discussed in Chapter 9.

Finally, we can consider **infinite graphs**, in which we drop the restriction that $V(G)$ and/or $E(G)$ are finite. If the number of edges incident with each vertex remains finite, then the infinite graph is said to be **locally finite**. Infinite graphs are discussed in **ST2**, Chapter 5.

Degrees

For each vertex v in a graph G, the number of edges incident to v is called the **degree** or **valency** of v, denoted by $d(v)$ or $\rho(v)$. The maximum and minimum degrees in G will be denoted by $\Delta(G)$ and $\delta(G)$, or by $\rho_{\max}$

and ρ_{min}. A vertex of degree 0 is called an **isolated vertex**, and a vertex of degree 1 is called an **end-vertex**.

The **degree-sequence** of G is the set of degrees of the vertices of G, usually arranged in non-decreasing order—for emphasis, this is sometimes called the **non-decreasing degree-sequence**; for example, the non-decreasing degree-sequence of the graph in Fig. 1 is (1, 2, 2, 3).

If all of the vertices of G have the same degree, then G is said to be a **regular graph**; if each degree is k, then G is a ***k*-regular** or ***k*-valent** graph. A 0-regular graph (that is, one with no edges) is called a **null graph**, and a 3-regular graph is often called a **trivalent graph**.

Analogous concepts can also be defined for digraphs. If v is a vertex of a digraph D, then its **in-degree** $\rho_{in}(v)$ (or indeg (v)) is the number of arcs in D of the form wv, and its **out-degree** or **score** $\rho_{out}(v)$ (or outdeg (v)) is the number of arcs in D of the form vw.

Subgraphs

A **subgraph** of a graph $G = (V(G), E(G))$ is a graph $H = (V(H), E(H))$ such that $V(H) \subseteq V(G)$ and $E(H) \subseteq E(G)$. If $V(H) = V(G)$, then H is called a **spanning subgraph** of G. If W is any set of vertices in G, then the **subgraph induced by *W*** is the subgraph of G obtained by taking the vertices in W and joining those pairs of vertices in W which are joined in G. An **induced subgraph** of G is a subgraph which is induced by some subset W of $V(G)$. Similar definitions may be given for digraphs and multigraphs.

If e is an edge of G, then the **edge-deleted subgraph** $G - e$ is the graph obtained from G by removing the edge e; more generally, we write $G - \{e_1, \ldots, e_k\}$ for the graph obtained from G by removing the edges $e_1, \ldots, e_k$. Similarly, if v is a vertex of G, then the **vertex-deleted subgraph** $G - v$ is the graph obtained from G by removing the vertex v together with all the edges incident to v; more generally, we write $G - \{v_1, \ldots, v_k\}$ for the graph obtained from G by removing the vertices $v_1, \ldots, v_k$ and all edges incident to any of them. These concepts are illustrated in Fig. 7.

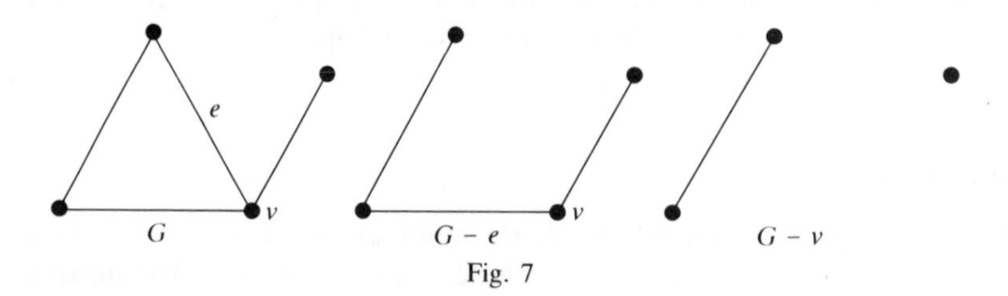

Fig. 7

New Graphs from Old

If G is a graph, then there are several graphs that we can obtain from G. For example, the **complement** of G (denoted by $\bar{G}$) is the graph with the same vertex-set as G, but where two vertices are adjacent if and only if they are not adjacent in G; a **self-complementary graph** is one which is isomorphic to its complement. The **line graph** $L(G)$ of G is the graph whose vertices correspond to the edges of G, and where two vertices are joined if and only if the corresponding edges of G are adjacent; line graphs are discussed in **ST1**, Chapter 10.

If $e = vw$ is an edge of G, then we can obtain a new graph by replacing e by two new edges vz and zw, where z is a new vertex; this is called **inserting a vertex into an edge** (see Fig. 8). If two graphs can be obtained

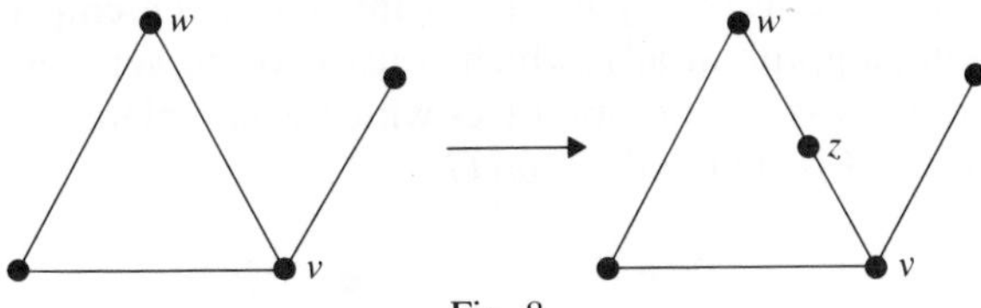

Fig. 8

from the same graph by inserting vertices into its edges, then these two graphs are called **homeomorphic**. We can also obtain a new graph from G by removing the edge $e = vw$ and identifying v and w in such a way that the resulting vertex is incident to all those edges (other than e) which were originally incident to v or to w; this is called **contracting the edge e** (see Fig. 9), and the resulting graph is denoted by G/e. If the graph H can be obtained from G by a succession of "edge-contractions" such as this, we say that G is **contractible** to H.

If G and G' are graphs with the same vertex-set, then their **intersection** $G \cap G'$ is the graph with edge-set $E(G) \cap E(G')$, and their **union** $G \cup G'$ is the graph with edge-set $E(G) \cup E(G')$. If G and G' are disjoint graphs, then $G \cup G'$ is the graph with vertex-set $V(G) \cup V(G')$ and edge-set $E(G) \cup E(G')$; it is sometimes called the **disjoint union** of G and G'. The disjoint union of k copies of G is often written kG. The **join** of G and G', denoted by $G + G'$, is obtained from their disjoint union by

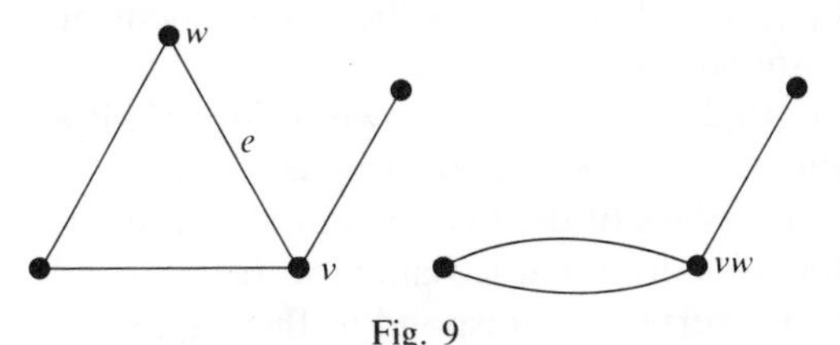

Fig. 9

adding edges joining each vertex of G to each vertex of G'. Finally, the **Cartesian product** of G and G' is the graph with vertex-set $V(G) \times V(G')$ in which the vertex (v, w) is adjacent to the vertex (v', w') whenever $v = v'$ and w is adjacent to w', or $w = w'$ and v is adjacent to v'.

Examples of Graphs

A graph in which every two vertices are adjacent is called a **complete graph**; the complete graph with p vertices and $\frac{1}{2}p(p-1)$ edges is denoted by K_p. The **circuit graph** of order p, denoted by C_p, consists of the vertices and edges of a p-gon, and the **path graph** P_p is obtained by removing an edge from C_p. The **null graph** N_p of order p is the graph with p vertices and no edges. The graphs K_5, C_5, P_5 and N_5 are shown in Fig. 10. It is also occasionally useful to introduce the **empty graph** (not strictly speaking a graph at all), which consists of no vertices or edges. A **clique** in a graph G is a subgraph of G which is complete; the size of the largest clique in G is denoted by $\omega(G)$.

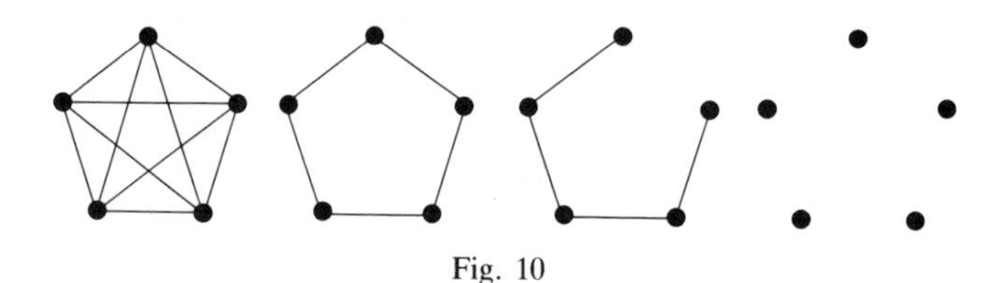

Fig. 10

A **bipartite graph** is one whose vertex-set can be partitioned into two sets (called **partite sets**) in such a way that each edge joins a vertex of the first set to a vertex of the second set. A **complete bipartite graph** is a bipartite graph in which every vertex in the first set is adjacent to every vertex in the second set; if the two partite sets contain r and s vertices, respectively, then the complete bipartite graph is denoted by $K_{r,s}$. Any complete bipartite graph of the form $K_{1,s}$ is called a **star graph**. A **complete multipartite** (or ***r*-partite**) **graph** is obtained by partitioning the vertex-set into r sets, and joining two vertices if and only if they lie in different sets; if all of these sets have size k, then the resulting graph is denoted by $K_{r(k)}$; note that $K_{r(k)}$ is the complement of rK_k. The graphs $K_{3,3}$ and $K_{3(3)}$ are shown in Fig. 11.

The **Petersen graph** is the graph shown in Fig. 12; it is the complement of the line graph of K_5. The **Platonic graphs** are the graphs corresponding to the vertices and edges of the five regular solids, the tetrahedron, cube, octahedron, dodecahedron and icosahedron (see Fig. 13). The ***k*-cube** Q_k is the graph whose vertices correspond to the sequences $(a_1, a_2, \ldots, a_k)$,

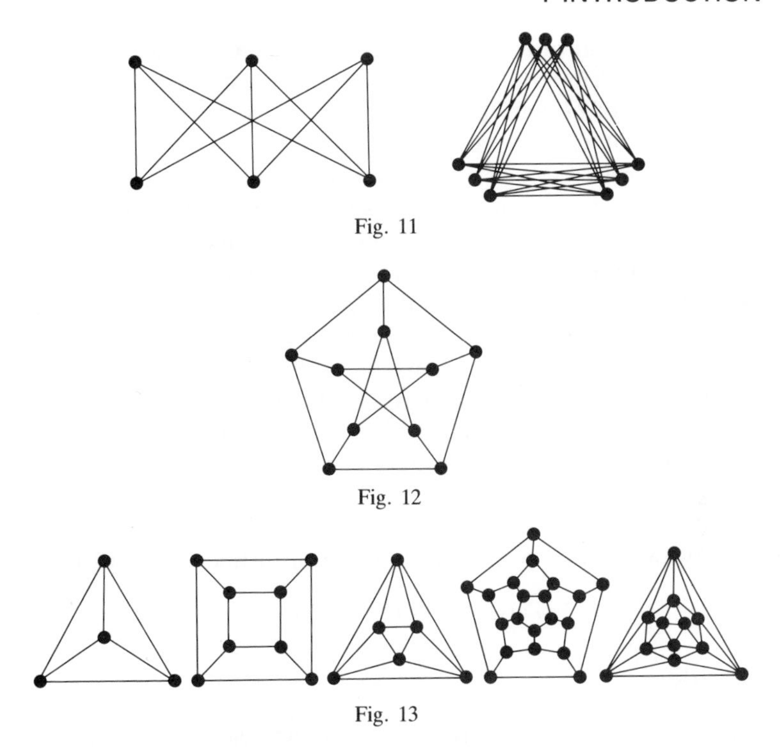

Fig. 11

Fig. 12

Fig. 13

where each $a_i = 0$ or 1, and whose edges join those pairs of vertices which correspond to sequences differing in just one place; thus $Q_2 = C_4$, and Q_3 is the graph of the cube.

Paths and Circuits

A sequence of edges of the form $v_0v_1, v_1v_2, \ldots, v_{r-1}v_r$ (sometimes abbreviated to $v_0v_1 \ldots v_r$) is called a **walk of length *r*** from v_0 to v_r; v_0 is called the **initial vertex** of the walk, and v_r is called the **terminal vertex**. If these edges are all distinct, then the walk is called a **trail**, and if the vertices $v_0, v_1, \ldots, v_r$ are also distinct, then the walk is called a **path**. Two paths in a graph are said to be **edge-disjoint** if they share no common edges; they are also said to be **vertex-disjoint** if they share no common vertices, although one frequently relaxes this condition to allow the initial vertices of the paths to coincide and also the terminal vertices. A walk or trail is said to be **open** if $v_0 \neq v_r$, and **closed** if $v_0 = v_r$; a walk in which the vertices $v_0, v_1, \ldots, v_r$ are all distinct except for v_0 and v_r (which coincide) is called a **circuit** (or **cycle**).

A circuit is said to be **even** if it has an even number of edges, and **odd** otherwise. A circuit of length 3 is called a **triangle**, and a circuit of length 4 is called a **quadrilateral**. The length of a shortest circuit in a graph G is called the **girth** of G, and the length of a longest circuit in G is called the **circumference** of G. If v and w are vertices in G, the length of any shortest path from v to w is called the **distance** between v and w, denoted by $d(v, w)$. The largest distance between two vertices in G is called the **diameter** of G; for example, the Petersen graph has diameter 2.

These definitions can be extended to directed graphs and infinite graphs. In particular, a **directed trail** in a digraph is a sequence of distinct arcs of the form $v_0v_1, v_1v_2, \ldots, v_{r-1}v_r$, a **directed path** is a sequence of arcs of the form $v_0v_1, v_1v_2, \ldots, v_{r-1}v_r$, where $v_0, v_1, \ldots, v_r$ are all distinct, and a **directed circuit** is a sequence of arcs of the form $v_0v_1, v_1v_2, \ldots, v_{r-1}v_0$, where $v_0, v_1, \ldots, v_{r-1}$ are all distinct. In an infinite graph, a **one-way infinite path** is a sequence of distinct edges of the form

$$v_0v_1, v_1v_2, \ldots, v_{r-1}v_r, \ldots \text{ or } \ldots, v_{-r}v_{-r+1}, \ldots, v_{-2}v_{-1}, v_{-1}v_0,$$

and a **two-way infinite path** is a sequence of distinct edges of the form

$$\ldots, v_{-r}v_{-r+1}, \ldots, v_{-1}v_0, v_0v_1, \ldots, v_{r-1}v_r, \ldots.$$

Connectivity

A graph G is **connected** if there is a path joining each pair of vertices of G (or, equivalently, if G cannot be expressed as the union of two disjoint graphs); a graph which is not connected is called **disconnected**. Clearly, every disconnected graph can be split up into a number of maximal connected subgraphs, and these subgraphs are called **components**. There are analogous definitions for digraphs; in particular, a digraph D is called **strongly connected** if, for each pair of vertices v and w, there is a directed path in D from v to w, and **connected** if there is a path from v to w in the underlying graph of D. Maximal strongly connected subgraphs are called **strong components**.

If G is a connected graph, and if the graph $G - v$ is disconnected, for some vertex v, then v is called a **cut-vertex** of G. More generally, a **separating set of vertices** in G is a set of vertices whose removal disconnects G. We say that a graph G with at least $k + 1$ vertices is ***k*-connected** if every two vertices v and w are connected by at least k paths which are pairwise disjoint except for the vertices v and w; a 2-connected graph is often called a **block** or a **non-separable graph**. The **connectivity** of G, denoted by $\kappa(G)$, is then defined to be the largest value of k for which G is k-connected.

If G is a connected graph, and if the graph $G - e$ is disconnected, for some edge e, then e is called a **bridge** of G. More generally, a **cutset** in G is a set of edges whose removal disconnects G. We say that a graph G is ***k*-edge-connected** if every two vertices v and w are connected by at least k edge-disjoint paths; the **edge-connectivity** of G, denoted by $\lambda(G)$, is then defined to be the largest value of k for which G is k-edge-connected. (Note that $\kappa(G) \leqslant \lambda(G)$.) Finally, we say that G is **cyclically *k*-edge-connected** if, by deleting fewer than k edges, we cannot disconnect G into components each of which contains a circuit.

The most important theorem relating these concepts is **Menger's theorem**; this takes several forms, among which are the following:

Theorem 1.1 (Menger's Theorem). *Let G be a connected graph ($\neq K_{k+1}$) with at least $k + 1$ vertices. Then*

(*i*) *G is k-connected if and only if G cannot be disconnected by the removal of $k - 1$ or fewer vertices;*

(*ii*) *G is k-edge-connected if and only if G cannot be disconnected by the removal of $k - 1$ or fewer edges.* ||

Further discussion of Menger's theorem and its many variations is given in **ST1**, Chapter 9, where the analogues for digraphs are also presented.

Finally, an ***r*-factor** in a graph G is an r-regular subgraph which includes every vertex of G. A necessary and sufficient condition for the existence of a 1-factor was given by Tutte [7]:

Theorem 1.2 (Tutte's Theorem). *A graph G has a* 1*-factor if and only if, for each set S of vertices of G, the subgraph $G - S$ has at most $|S|$ components of odd order.* ||

Traversability

A connected graph G is **Eulerian** if it has a closed trail which includes every edge of $E(G)$; such a trail is called an **Eulerian trail**. Similarly, a strongly connected digraph D is **Eulerian** if it has a closed directed trail which includes every arc of $A(D)$. Necessary and sufficient conditions for a graph or digraph to be Eulerian are given in the following theorem:

Theorem 1.3. (*i*) *A connected graph G is Eulerian if and only if every vertex of G has even degree;*

(*ii*) *a strongly connected digraph D is Eulerian if and only if the in-degree and out-degree of each vertex are equal.* ||

Eulerian graphs are discussed in **ST2**, Chapter 2. Note that the definitions presented there are slightly different from those just given.

A graph G is **Hamiltonian** if it has a circuit which includes every vertex of $V(G)$; such a circuit is called a **Hamiltonian circuit**. More generally, a graph G is **traceable** if it has a path which includes every vertex of $V(G)$; such a path is called a **Hamiltonian path**. Analogous definitions can be given for digraphs, and the theory of Hamiltonian graphs and digraphs is discussed in **ST1**, Chapter 6.

Trees

A connected graph which contains no circuits is called a **tree**, and a graph whose components are trees is called a **forest**, or **acyclic graph**. The trees of order 5 are shown in Fig. 14.

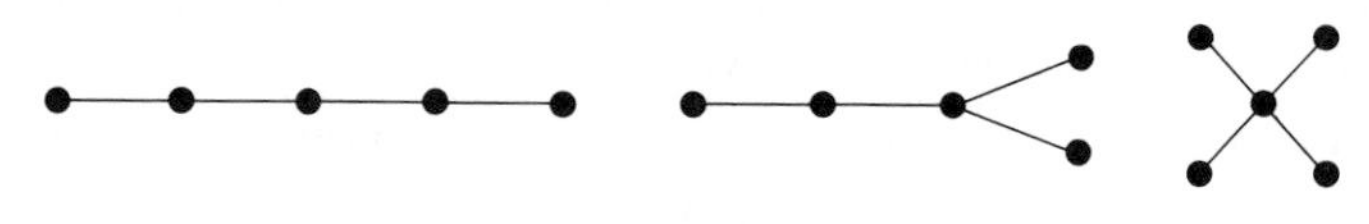

Fig. 14

The main properties of trees are summarized in the following theorem:

Theorem 1.4. *If T is a tree of order p, then*

(*i*) *T is a connected graph with $p - 1$ edges*;

(*ii*) *every edge of T is a bridge*;

(*iii*) *if v is a vertex of T with $d(v) > 1$, then v is a cut-vertex*;

(*iv*) *if v and w are distinct vertices of T, then there is exactly one path from v to w.* ||

If G is a connected graph, then a **spanning tree** in G is a connected spanning subgraph containing no circuits. A method for calculating the number of spanning trees in a given graph is described later in this chapter. The **arboricity** of G is the minimum number of forests whose union is G.

Planar Graphs

A **planar graph** is a graph which can be embedded in the plane in such a way that no two edges intersect geometrically except at a vertex to which they are both incident. A graph embedded in the plane in this way is called a **plane graph**; in this case, the points of the plane not on G are partitioned into open sets called **faces**, or **regions** (see Fig. 15), and the number r of such faces is given by **Euler's polyhedron formula**:

Theorem 1.5 (Euler's Polyhedron Formula). *Let G be a connected plane graph with $p(\geqslant 3)$ vertices, q edges and r faces. Then*

$$p - q + r = 2. \;\|$$

Since G has no loops or multiple edges, every face must be bounded by at least three edges. It follows that $2q \leqslant 3r$, and hence that $q \leqslant 3p - 6$; equality holds when every face is bounded by a triangle, and such a graph is called a **triangulation**. If G contains no triangles, then $2q \leqslant 4r$, and so $q \leqslant 2p - 4$. It is a simple matter to check that every planar graph contains a vertex of degree 5 or less, and has girth at most 5. A connected plane graph which contains no bridges is sometimes called a **map**, and a planar graph which can be embedded in the plane in such a way that every vertex lies on the boundary of the same face is called an **outerplanar graph**; an **outerplane graph** is defined analogously.

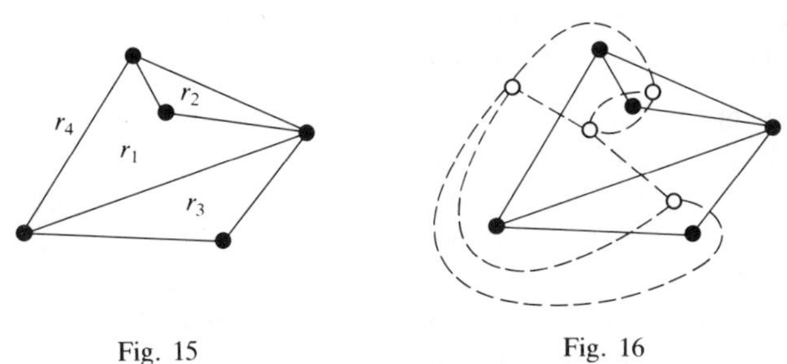

Fig. 15 Fig. 16

A necessary and sufficient condition for a graph to be planar has been given by Kuratowski [**6**]; we present two forms of this result:

Theorem 1.6 (Kuratowski's Theorem). *A graph G is planar if and only if*

either (*i*) *G has no subgraph homeomorphic to K_5 or $K_{3,3}$;*

or (*ii*) *G has no subgraph contractible to K_5 or $K_{3,3}$.* $\|$

If G is a connected plane graph, then its **dual graph** G^* is the (general) graph obtained by the following procedure:

(*i*) place a point inside each face of G—these points correspond to the vertices of G^*;

(*ii*) for each edge e of G, draw a line joining the vertices in the two faces bounded by e—these lines correspond to the edges of G^* (see Fig. 16).

It is easy to see that G^* is a plane graph whose dual graph is isomorphic to G, and that if G has p vertices, q edges and r faces, then G^* has r vertices, q edges and p faces.

The Coloring of Graphs

If G is a graph, we define its **chromatic number** $\chi(G)$ to be the minimum number of colors needed to color the vertices of G in such a way that no two adjacent vertices are assigned the same color. If $\chi(G) = k$, we say that G is **k-chromatic**, and if $\chi(G) \leqslant k$, we say that G is **k-colorable**. For example, the complete graph K_p is p-chromatic, the path graph $P_p (p \geqslant 2)$ is 2-chromatic, and the circuit graph C_p is 2-chromatic or 3-chromatic according to whether p is even or odd. Note that if G is a bipartite graph, then G is 2-colorable. The set of all vertices with the same color is called a **color class**.

An upper bound for the chromatic number of a graph G has been given by Brooks [**3**], and involves the maximum degree of G:

Theorem 1.7 (Brooks' Theorem). *Let G be a connected graph which is not a complete graph or a circuit of odd length, and let ρ be the largest degree in G. Then G is ρ-colorable.* ||

For each graph G, let $P_G(k)$ be the number of ways of coloring the vertices of G in such a way that no two adjacent vertices are assigned the same color. For example, if $G = K_p$, then $P_G(k) = k(k-1) \ldots (k-p+1)$, and if $G = P_p$, then $P_G(k) = k(k-1)^{p-1}$. It is not difficult to show that if G has p vertices and q edges, then $P_G(k)$ is a monic polynomial in k of degree p, in which the coefficients alternate in sign, the constant coefficient is zero, and the coefficient of k^{p-1} is $-q$; $P_G(k)$ is called the **chromatic polynomial** of G (see Chapter 2).

We also define the **chromatic index** $\chi'(G)$ of G to be the minimum number of colors needed to color the edges of G in such a way that no two adjacent edges are assigned the same color. If $\chi'(G) = k$, we say that G is **k-edge-chromatic**, and if $\chi'(G) \leqslant k$, we say that G is **k-edge-colorable**. Sharp bounds for $\chi'(G)$ were obtained by Vizing [**8**]:

Theorem 1.8 (Vizing's Theorem). *If G has maximum degree Δ, then*

$$\Delta \leqslant \chi'(G) \leqslant \Delta + 1. \quad ||$$

We end this section with the famous **four-color theorem** for planar graphs, discussed at some length in **ST1**, Chapters 4 and 15:

Theorem 1.9 (Four-Color Theorem). *Every planar graph is four-colorable.* ||

Matrices

If G is a graph with vertex-set $\{v_1, v_2, \ldots, v_p\}$, then the **adjacency matrix** of G is the $p \times p$ matrix $\mathbf{A}(G) = (a_{ij})$, where

$$a_{ij} = \begin{cases} 1, & \text{if } v_i \text{ and } v_j \text{ are adjacent,} \\ 0, & \text{if not.} \end{cases}$$

The adjacency matrix of a digraph is defined similarly.

In **ST1**, Chapter 11 we studied the eigenvalues of $\mathbf{A}(G)$, which are clearly independent of the way in which the vertices are labeled. For convenience, we refer to these eigenvalues as the **eigenvalues of *G***, and in a similar way, the characteristic polynomial of $\mathbf{A}(G)$ is called the **characteristic polynomial of *G***.

There are various other matrices associated with G. For example, if G has edge-set $\{e_1, e_2, \ldots, e_q\}$, then the **incidence matrix** of G is the $p \times q$ matrix $\mathbf{B}(G) = (b_{ij})$, where

$$b_{ij} = \begin{cases} 1, & \text{if } v_i \text{ is incident to } e_j, \\ 0, & \text{if not.} \end{cases}$$

An example of the use of matrices in graph theory is provided by the **matrix-tree theorem** (see, for example, [**5**]):

Theorem 1.10 (Matrix-tree Theorem). *Let G be a connected labeled graph with adjacency matrix $\mathbf{A}$, and let $\mathbf{M}$ be the matrix obtained from $-\mathbf{A}$ by replacing each diagonal entry by the degree of the corresponding vertex. Then all of the cofactors of $\mathbf{M}$ are equal, and their common value is the number of spanning trees in G.* ||

The Efficiency of Algorithms

In several chapters of this book we consider the efficiency of various graph algorithms. For each graph G under consideration we associate a number n called its *size*—n is usually the order or the number of edges of G.

The most efficient algorithms are **polynomial-time algorithms**, which can be solved in time cn^k, for some constants c and k. The class of problems which can be solved in polynomial time is denoted by P. A problem is called an ***NP*-problem** if its solution, when given, can be checked in polynomial time, even if it took exponential time to find the solution originally. It is clear that $P \subseteq NP$, where NP denotes the class of all NP-problems. It is unknown whether $P = NP$, although it is generally believed that this is *not* the case.

Finally, there is a class of important problems known as ***NP*-complete problems**. These problems, which include the traveling salesman problem, the graph isomorphism problem, the Hamiltonian circuit problem (is G Hamiltonian?), and the k-colorability problem (is G k-colorable?), have the property that if any one of them is in P, then so are all of them. It is generally believed that *none* of them lies in P. Further information on this topic is given in [**4**].

And finally . . .

If S is a finite set, we denote the number of elements in S by $|S|$; the empty set is denoted by $\emptyset$. We use $\lfloor x \rfloor$ for the largest integer not greater than x, and $\lceil x \rceil$ for the smallest integer not smaller than x (so that, for example, $\lfloor \pi \rfloor = 3$ and $\lceil \pi \rceil = 4$). The sets of real numbers, rationals, positive integers, and integers are denoted by **R**, **Q**, **N** and **Z**. As in this chapter, the end or absence of a proof is denoted by ||. In the references for each chapter, *Mathematical Reviews* numbers are indicated by *MR***15**–234 (page 234 of Volume **15**), *MR***35**#234 (review number 234 of Volume **35**), or *MR***82a**:05047 (review number 47 of Section 05 of Issue **82a**).

References

1. C. Berge, *Graphs and Hypergraphs* (transl. E. Minieka), North-Holland, Amsterdam, 1973; *MR***50**#9640.
2. J. A. Bondy and U. S. R. Murty, *Graph Theory with Applications*, American Elsevier, New York, and MacMillan, London, 1976; *MR***54**#117.
3. R. L. Brooks, On colouring the nodes of a network, *Proc. Cambridge Phil. Soc.* **37** (1941), 194–197; *MR***6**–281.
4. M. Garey and D. S. Johnson, *Computers and Intractability. A Guide to the Theory of NP-Completeness*, W. H. Freeman, San Francisco, 1979; *MR***80g**:68056.
5. F. Harary, *Graph Theory*, Addison-Wesley, Reading, Mass., 1969; *MR***41**#1566.
6. K. Kuratowski, Sur le problème des courbes gauches en topologie, *Fund. Math.* **15** (1930), 271–283.
7. W. T. Tutte, The factorization of linear graphs, *J. London Math. Soc.* **22** (1947), 107–111; *MR***9**–297d.
8. V. G. Vizing, On an estimate of the chromatic class of a p-graph (Russian), *Diskret. Analiz* **3** (1964), 25–30; *MR***31**#4740.
9. R. J. Wilson, *Introduction to Graph Theory*, 3d ed., Longman Group, Harlow, Essex, 1985; *MR***50**#9643, **80g**:05028.

2
Chromatic Polynomials

R. C. READ and W. T. TUTTE[†]

1. Introduction

Suppose we are given a graph G and a set Λ of λ objects called "colors". We define a **λ-coloring** of G as a mapping of the vertex-set $V(G)$ of G into the color-set Λ, subject to the restriction that each edge of G must join vertices of two different colors. We denote the number of λ-colorings of G by $P(G, \lambda)$, or simply by $P(\lambda)$ if there is no likelihood of confusion.

This function $P(G, \lambda)$—which is actually a polynomial—was introduced by G. D. Birkhoff in 1912 [**5**]. He hoped that it would help to prove the *four-color theorem*, which can be stated in the form

$$P(G, 4) > 0, \textit{ for any loopless planar graph } G.$$

So far, this hope has not been borne out, but the chromatic function seems, nevertheless, to be of great interest in its own right. The *fons et origo* of the extensive study of this chromatic function is the remarkable paper by G. D. Birkhoff and D. C. Lewis [**7**], which appeared in 1946 and is still recommended reading for would-be researchers. This paper, and much early work on the topic, was concerned mostly with planar graphs; indeed, Birkhoff defined what we now call the **chromatic polynomial** of a planar graph in terms of its dual map. The notion was extended to graphs in general by Hassler Whitney [**32**], [**33**].

[†]This research was supported by the National Sciences and Engineering Research Council of Canada under grants A8142 and A3113.

GRAPH THEORY, 3
ISBN 0-12-086203-4

We give, in the next section, a proof that $P(G, \lambda)$ is a polynomial in λ. Once this is agreed, there is no need to restrict λ to be a non-negative integer, and there is now much interest in the properties of $P(G, \lambda)$ for certain special non-integer values of λ (see Sections 4 and 5). Even for complex values of λ, there are some interesting facts of observation, although there are as yet no theorems.

In other sections we discuss the problem of how to compute chromatic polynomials as efficiently as possible. We also look at problems concerning the extent to which a chromatic polynomial determines its graph.

Before beginning our development, we make a couple of observations about loops and multiple edges. First, a graph G with a loop clearly has no coloring for any λ, so that we must have $P(G, \lambda) = 0$. Hence, we shall assume that all graphs are loopless. Secondly, we note that if two vertices are adjacent, then the number of edges joining them does not affect the number of colorings. Therefore, we shall generally assume that no multiple edges are present. However, in the case of plane triangulations, which we consider in Section 5, it will be convenient to allow such edges.

2. Basic Properties of Chromatic Polynomials

The first thing we need to show is that $P(G, \lambda)$ is indeed a polynomial.

Theorem 2.1. *If G is a graph on p vertices, then $P(G, \lambda)$ is a polynomial of degree p.*

Proof. For each non-negative integer k, let $\alpha(G, k)$ denote the number of partitions of $V(G)$ into exactly k non-empty subsets, such that no edge of G joins two vertices in the same subset. From a set of λ colors, there are

$$\lambda(\lambda - 1)(\lambda - 2) \ldots (\lambda - k + 1)$$

ways of allocating a different color to each of the subsets, and each of these gives a coloring of G. It follows that

$$P(G, \lambda) = \sum_k \lambda(\lambda - 1)(\lambda - 2) \ldots (\lambda - k + 1)\alpha(G, k),$$

which is clearly a polynomial of degree p. ||

A useful fact about polynomials is that if two polynomials agree for infinitely many values of the variable (such as, for example, all positive integers), then they must be identical. We shall frequently use this in establishing polynomial identities.

We now give several easily proved results on chromatic polynomials:

Theorem 2.2. *The edge-free graph N_p of order p has chromatic polynomial*

$$P(N_p, \lambda) = \lambda^p. \ ||$$

Theorem 2.3. *The complete graph K_p of order p has chromatic polynomial*

$$P(K_p, \lambda) = \lambda(\lambda - 1)(\lambda - 2) \ldots (\lambda - p + 1). \;\|$$

Theorem 2.4. *If a graph G is obtained from a graph H by adjoining a new vertex adjacent to every vertex of H, then*

$$P(G, \lambda) = \lambda P(H, \lambda - 1).$$

Proof. This is true for any positive integer λ since, having chosen a color for the new vertex (λ possibilities), we must color the original graph with the remaining $\lambda - 1$ colors. ||

To the foregoing results we add two *Propositions of Recursion*. These will be extremely useful in later sections.

Theorem 2.5. *Let G be the union of two subgraphs G_1 and G_2, whose intersection is a complete graph of order k. Then*

$$P(G, \lambda) = \frac{P(G_1, \lambda) \cdot P(G_2, \lambda)}{\lambda(\lambda - 1) \ldots (\lambda - k + 1)}.$$

Proof. To establish this result for an arbitrary positive integer λ, we note that any coloring of G can be obtained by combining a coloring of G_1 with any coloring of G_2 which preserves the colors given to the vertices in the intersection. This restriction has the effect of reducing the number of colorings of G_2 by a factor of $\lambda(\lambda - 1)(\lambda - 2) \ldots (\lambda - k + 1)$, from which the result follows. ||

We note that this result can be regarded as including the case $k = 0$, when G_1 and G_2 are disjoint.

Before stating our second recursion result, we recall that, for an edge e in a graph G, $G - e$ and G/e denote, respectively, the graphs obtained by deleting and contracting e. (In the contraction, we assume that there is no loop, and that double edges are replaced by single ones.)

Theorem 2.6 (The Deletion–Contraction Formula). *Let G be a graph and e an edge of G. Then*

$$P(G, \lambda) = P(G - e, \lambda) - P(G/e, \lambda).$$

Proof. Note that colorings of $G - e$ are of two kinds: those in which the ends of e receive different colors, and those in which they receive the same color. The former correspond in a one-to-one manner with the colorings of G, and the latter with the colorings of G/e. Thus the equation holds for all positive integer values of λ, and consequently for all λ. ||

The recursion formulas of Theorems 2.5 and 2.6 can be used in the construction of tables of chromatic polynomials, for they enable us to express the chromatic polynomial of any graph in terms of the chromatic polynomials of smaller graphs. This is illustrated in Figs. 1 and 2. Note that in these figures we have used a handy notational device introduced by Zykov [**36**], which is to let a drawing of a graph represent its chromatic polynomial (with the variable λ understood).

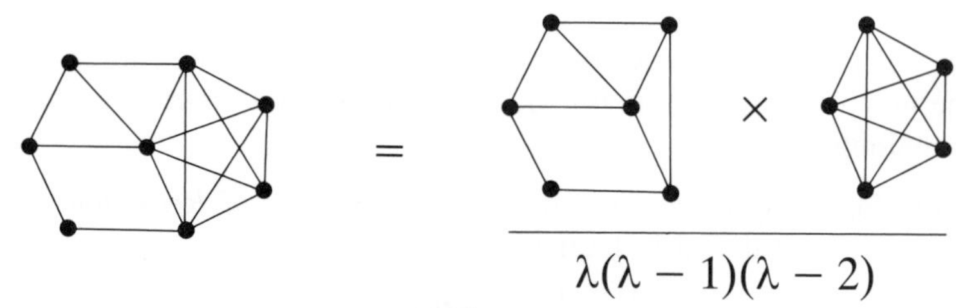

Fig. 1

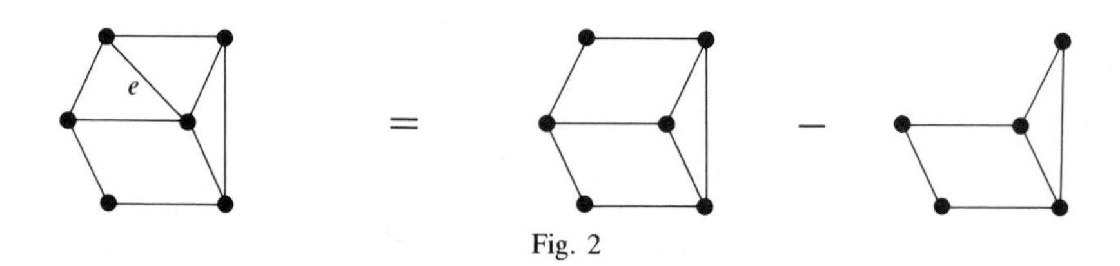

Fig. 2

These same two results are often used in inductive proofs of theorems about chromatic polynomials. Consider, for example, the following theorem:

Theorem 2.7. *Let G be a graph with p vertices, q edges and k components. Anticipating result (iii) below, we denote the coefficient of* λ^j *in* $P(G, \lambda)$ *by* $(-1)^{p-j}a_j$. *Then*

(*i*) a_j *is non-zero if and only if* $k \leq j \leq p$;

(*ii*) $a_p = 1$ *and* $a_{p-1} = q$;

(*iii*) a_j *is positive for* $k \leq j \leq p$.

Proof. We use induction on the number q of edges. If $q = 0$, the theorem is true, by Theorem 2.2. Otherwise we choose an edge e and use Theorem 2.6. By the induction hypothesis the results hold for both $P(G - e, \lambda)$ and $P(G/e, \lambda)$. Since G/e has one fewer vertex than $G - e$ and G, and has the same number of components as G, while $G - e$ has the same number or one more, the results must also hold for G. ||

Corollary 2.8. *For every graph G,* $P(G, \lambda)$ *is a monic polynomial whose coefficients alternate in sign.* ||

We note here that Whitney [32] has given a formula for a_j as a sum over all subgraphs of G of a particular kind. Suppose that the edges of G have been labeled arbitrarily with the integers 1, 2, ..., q. Consider all the circuits of G, and from each one remove the edge with greatest label, giving what can be called a *broken circuit*. Then Whitney's result is that

a_j = *the number of subgraphs of* G, *having* $p - j$ *edges, which do not contain any broken circuit.*

Result (*ii*) above follows at once from this, and so does

$$a_{p-2} = \binom{q}{2} - N_T,$$

where N_T is the number of triangles in G. This is a special case of the following theorem (Meredith [23]) on the **girth** (the length of a shortest circuit) of G:

Theorem 2.9. *If* g *is the girth of* G *and* k_g *is the number of circuits of length* g, *then*

$$a_{p-j} = \binom{q}{j} \text{ for } j < g - 1,$$

$$a_{p-g+1} = \binom{q}{g-1} - k_g,$$

and

$$a_{p-j} < \binom{q}{j} \text{ for } j > g - 1. \quad \|$$

This, in turn, 's a special case of the following result of Li and Tian [20]:

Theorem 2.10. *If* g *is the girth of* G *and* k_g *is the number of circuits of length* g, *then*

$$a_{p-j} \leq \binom{q}{j} - \binom{q-g+2}{j-g+2} + \binom{q-k_g-g+2}{j-g+2}. \quad \|$$

These same authors also give the following rather curious inequality which goes the other way:

Theorem 2.11. *If* k_n *is the number of circuits of length* n *in* G, *and* $3 \leq n \leq l \leq p$, *then*

$$a_r \geq \sum_{i=0}^{l} \binom{q-p+i}{i}\binom{p-1-i}{r-1} - \sum_{n=3}^{l} \left\{ k_n \sum_{i=0}^{l-n} \binom{q-p+i}{i}\binom{p-n-i}{r-1} \right\}. \quad \|$$

Other inequalities for the chromatic coefficients are known. It can be shown (for example, using Whitney's broken circuit interpretation) that for any given graph the absolute values of the coefficients are the simplex counts for a certain simplicial complex. As such, they obey various known inequalities for such counts. For further information and details on this way of looking at the coefficients, see Wilf [**34**].

We now state a very 'obvious' property of the coefficients a_j:

The Unimodal Conjecture. *For any chromatic polynomial, the statement*

$$a_j > a_{j+1} \text{ and } a_{j+1} < a_{j+2}$$

is false for every j.

In other words, this conjecture asserts that the coefficients first increase in absolute value, and then decrease. This behavior of the coefficients is very apparent if one looks at lists of chromatic polynomials, but there is as yet no proof of this unimodal property.

3. Some Families of Graphs

Thus far, the chromatic polynomials of only two families of graphs have been given: those of the complete graphs, and of their complements, the edge-free graphs. In this section, we use the recursion formulas of Section 2 to establish a few more results, beginning with trees.

For the graph K_1 we have $P(G, \lambda) = \lambda$, while for the graph K_2 we have $P(G, \lambda) = \lambda(\lambda - 1)$, by Theorem 2.3. From these results, by repeated application of Theorem 2.5 with $k = 1$, we obtain the following theorem:

Theorem 3.1. *The chromatic polynomial of any tree* T *of order* p *is*

$$P(T, \lambda) = \lambda(\lambda - 1)^{p-1}. \quad \|$$

We next consider the family of circuits:

Theorem 3.2. *For* $p \geq 3$, *the chromatic polynomial of the p-circuit* C_p *is*

$$P(C_p, \lambda) = (\lambda - 1)^p + (-1)^p(\lambda - 1).$$

Proof. We use induction on the order p. We have, by Theorem 2.3,

$$P(C_3, \lambda) = \lambda(\lambda - 1)(\lambda - 2),$$

which is the required expression when $p = 3$. If e is an edge of C_p, we have, by Theorems 2.6 and 3.1,

$$P(C_p, \lambda) = \lambda(\lambda - 1)^{p-1} - P(C_{p-1}, \lambda).$$

The induction is now easily completed. ||

We note that, for $p = 1$ and 2, this result gives the chromatic polynomials for the graph consisting of a single loop and for the 'link graph' K_2, respectively.

The **wheel** W_p of order p (which can also be considered as a pyramid) consists of the circuit C_{p-1} with an additional vertex joined to all the others. If we similarly adjoin another vertex adjacent to all those in the wheel W_{p-1}, we obtain a graph which it seems natural to call a **biwheel** U_p. The **bipyramid** B_p of order p is obtained from U_p by deleting the edge joining the two added vertices. We note that U_5 is K_5 and that B_6 is the graph of the octahedron. The graph W_9 is shown in Fig. 3.

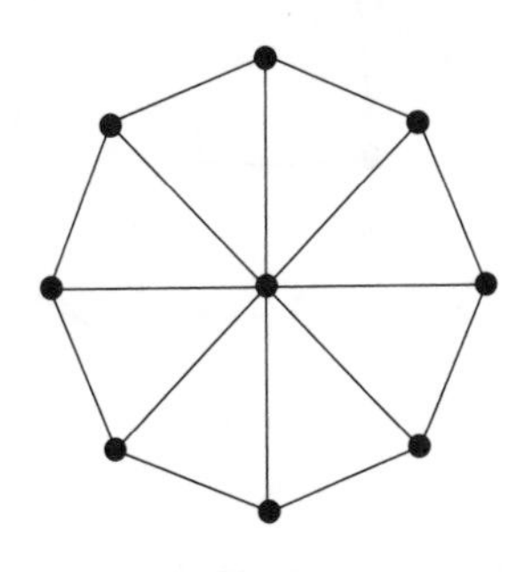

Fig. 3

Theorem 3.3.

(*i*) $P(W_p, \lambda) = \lambda\{(\lambda - 2)^{p-1} + (-1)^{p-1}(\lambda - 2)\}$;

(*ii*) $P(U_p, \lambda) = \lambda(\lambda - 1)\{(\lambda - 3)^{p-2} + (-1)^p(\lambda - 3)\}$;

(*iii*) $P(B_p, \lambda) = \lambda\{(\lambda - 2)^{p-2} + (\lambda - 1)(\lambda - 3)^{p-2} + (-1)^p(\lambda^2 - 3\lambda + 1)\}$.

Proof. First, (*i*) follows from Theorems 2.4 and 3.2, while (*ii*) follows from (*i*) and Theorem 2.4. Finally, we observe that if e is the edge joining the two vertices of maximum degree in $U_p (p \geq 6)$, the $U_p - e = B_p$ and $U_p/e = W_{p-1}$. Therefore, by Theorem 2.6,

$$P(B_p, \lambda) = P(U_p, \lambda) + P(W_{p-1}, \lambda),$$

from which (*iii*) follows. ||

We note here for future reference that B_p is a planar graph. We can represent the circuit C_{p-2} by the equator of a sphere, and the other two vertices by the North and South poles. The edges joining these two vertices to the vertices of C_{p-2} are then 90° arcs of great circles (see Fig. 4). This representation of B_p defines a map on the sphere having $2p - 4$

faces, or regions, each of which is triangular. By stereographic projection to a plane from any general point of the sphere, we obtain a triangulation of the plane, or, equivalently, a planar drawing of the graph B_p.

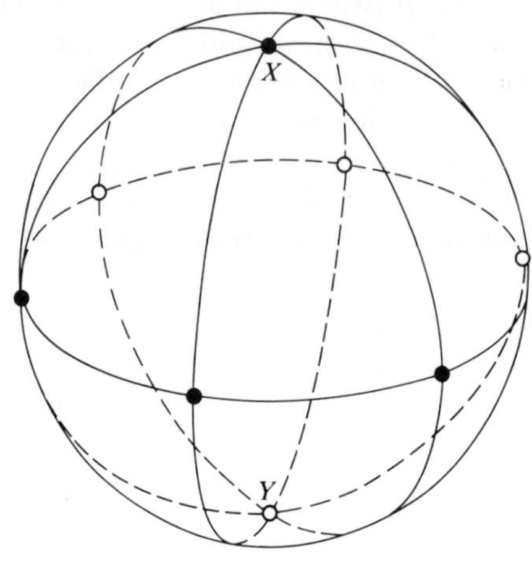

Fig. 4

4. Zeros of Chromatic Polynomials

The chromatic polynomial of any graph is zero at $\lambda = 0$, and for any graph with an edge there must be a zero of $P(G, \lambda)$ at $\lambda = 1$. Furthermore, any graph containing an odd circuit must have a zero at $\lambda = 2$. The reason is that if $P(G, \lambda)$ has a zero at the positive integer k, then G has no k-coloring. Accordingly, $P(G, \lambda)$ then has a zero at each non-negative integer less than k.

It is usual to find that $P(G, \lambda)$ also has one or more non-integer real zeros. It is convenient to note here the absence of zeros in certain parts of the real line.

Theorem 4.1. *Let G be a graph of order p, and λ a negative real number. Then $P(G, \lambda)$ is non-zero and has the sign of $(-1)^p$.*

Proof. This theorem is an immediate consequence of Theorem 2.8. ||

Theorem 4.2. *Let G be a connected graph with at least one edge, and let λ be a real number satisfying $0 < \lambda < 1$. Then $P(G, \lambda)$ is non-zero, and it has the sign of $(-1)^{p+1}$.*

Proof. We note first that this result is true when G has only one edge, by Theorem 2.4, for then $P(G, \lambda) = \lambda(\lambda - 1)$. Suppose that H is a graph for which the theorem fails and which has the least number of edges, q say, consistent with this condition. Then $q \geq 2$.

Choose an edge e. If $G - e$ is not connected, then e is a cut-edge. We can now deduce the theorem for G from its truth for smaller graphs, by using Theorem 2.5 with $k = 1$. If $G - e$ is connected, we use Theorem 2.6 instead, noting that $P(G/e, \lambda)$ satisfies the theorem. ||

This theory can be taken a little further. It can be shown that if G is connected, then its zero at $\lambda = 0$ has multiplicity 1. Moreover, if G is non-separable, then the zero at $\lambda = 1$ also has multiplicity 1.

Sometimes Theorem 4.2 is exploited in connection with the properties of algebraic integers. For example, let us write

$$\tau = \tfrac{1}{2}(1 + \sqrt{5}) \text{ and } \tau^* = \tfrac{1}{2}(1 - \sqrt{5}).$$

Then τ and τ^* are the roots of the quadratic equation

$$x^2 = x + 1.$$

By the proof of Theorem 2.1, $P(G, \lambda)$ is a polynomial in λ with integral coefficients. Hence if it is zero at $\lambda = 1 + \tau$, it must also be zero at the conjugate number $1 + \tau^*$. But $1 + \tau^*$ lies between 0 and 1. We can therefore deduce the following from Theorem 4.2; we may, of course, write τ^2 instead of $\tau + 1$:

Theorem 4.3. *Let G be a connected graph with at least one edge. Then $P(G, \tau + 1)$ is non-zero.* ||

For some further results on real roots of chromatic polynomials, see Woodall [**35**].

When we come to consider complex roots of chromatic polynomials, we get some tantalizing glimpses of possible theorems, but very little in the way of concrete results. It has been found, for example, that when the roots of the chromatic polynomials of all the graphs of some specific class of graphs are plotted in the complex plane, it frequently happens that they show a very non-random distribution, often tending to cluster near to certain fairly well delineated smooth curves. Several examples of this behavior are discussed in [**4**], but the reasons for it are not well understood.

In **ST1**, Chapter 15, one of us drew attention to a curious phenomenon that is evident when the roots of the chromatic polynomials of all the graphs on eight vertices are plotted in the complex plane—namely, that the roots tend to avoid the immediate neighborhood of certain 'popular' roots. In consequence, these roots are surrounded by circles within which no other roots occur. No explanation for this behavior has yet been forthcoming, but possibly related to it are some results contained in a thesis by Thier [**27**], which define regions of the complex plane within which the roots must lie. We have not seen a copy of this thesis, but the results, as quoted by Gernert in [**16**], are as follows:

Theorem 4.4. *All zeros of a chromatic polynomial $P(G, z)$ lie within the following closed regions of the complex plane:*

(*i*) *the union of the circular disks $|z - q| \leq q$ and $|z| \leq q - 1$;*

(*ii*) *the union of the circular disks $|z - q + p - 2| \leq q$ and $|z - 1| \leq q - 1$;*

(*iii*) *the union of the circular disk $|z - 1| \leq q - 1$ and the Cassini oval $|z - q + p - 2| \cdot |z - 1| \leq q(q - 1)$.* ||

5. Triangulations

Let a connected graph G be drawn on the sphere so as to give a map T in which each face is triangular. Such a map is called a **triangulation of the sphere**. We say that the chromatic polynomial $P(G, \lambda)$ of G is also the *chromatic polynomial* $P(T, \lambda)$ of T. We saw in Section 3 that the bipyramid B_p can be drawn on the sphere so as to give a map of this kind. The simplest of all triangulations has three vertices, three edges and two faces. For convenience in inductive proofs we shall allow a triangulation to have *digons*—that is, pairs of edges with the same ends. Such a digon must not bound a face, and both its interior and exterior must be triangulated. A triangulation with a digon can thus be decomposed into two smaller triangulations by cutting along the digon, and then closing up the digon in each piece. The number of faces of a triangulation T is always even, and the chromatic polynomial $P(T, \lambda)$ vanishes at $\lambda = 0$, $\lambda = 1$ and $\lambda = 2$. Furthermore, it is easy to show that

$$P(T, 3) = 0 \text{ unless } T \text{ is Eulerian.}$$

If T is Eulerian, then there is essentially only one way of 3-coloring its graph, apart from permutations of the colors, so that (allowing such permutations) we have $P(T, 3) = 6$.

The proof of the four-color theorem **[2]**, **[3]** is of interest since it allows us to assert that $P(T, 4) > 0$. Earlier work tells us that $P(T, \lambda) > 0$ for $\lambda \geq 5$, but it is still not known whether zeros can occur between 4 and 5.

When the chromatic zeros of a triangulation are calculated, it is usually found that there is one near $\tau^2 = \frac{1}{2}(3 + \sqrt{5})$. This observation has led to a family of theorems on values of chromatic polynomials when $\lambda = \tau^2$. The first which we give here, the so-called *Vertex Elimination Theorem*, is concerned with plane graphs having certain wheel-like subgraphs:

Theorem 5.1 (Vertex Elimination Theorem). *Let G be a plane graph with a wheel subgraph W consisting of a vertex v and its m neighbors in G. Then*

$$P(G, \tau^2) = (-1)^m \tau^{1-m} P(G - v, \tau^2).$$

Proof. The proof is by induction on the number r of vertices and edges of G not in W—that is, $r = p + q - (3m + 1)$.

If $r = 0$, then the result follows from the known formulas for the chromatic polynomials of circuits and wheels. The step from $r = n$ to $r = n + 1$ in the induction is easily effected by the recursion formulas of Theorems 2.5 and 2.6. ||

The following result, due to Tutte **[28]**, shows that the chromatic polynomial of a triangulation is small near τ^2, so that it seems reasonable that there will be a zero nearby:

Theorem 5.2. *If T is a triangulation with p vertices, then*

$$|P(T, \tau^2)| \leq \tau^{5-p}.$$

Proof. For the proof we use induction on p. It is trivial when p has its smallest value, 3. The step from p to $p + 1$ is effected by means of Theorem 2.5 when G has a digon. (Note that for triangulations, double edges are not trivial if they form a digon.) In the remaining case, we apply Theorem 5.1 to a vertex v of G with smallest degree, and then use the recursion formulas to express $P(G, \tau^2)$ in terms of the chromatic polynomials of triangulations with at most p vertices. ||

A third member of this family of theorems is concerned with a plane graph G in which there is a 4-circuit $C = xyztx$, with no edge or vertex of G in its interior except for one diagonal $e = xz$. As for Theorem 5.1, G need not be a triangulation, or even connected. We now define θ_e as the operation of 'twisting' A inside C—that is, replacing it by the diagonal yt. We write φ_e for the operation on G of contracting xz to a single new vertex x' and deleting one member of each resulting pair of double edges. A similar operation on $\theta_e G$, contracting yt, gives us a graph $\psi_e(G)$ (see Fig. 5).

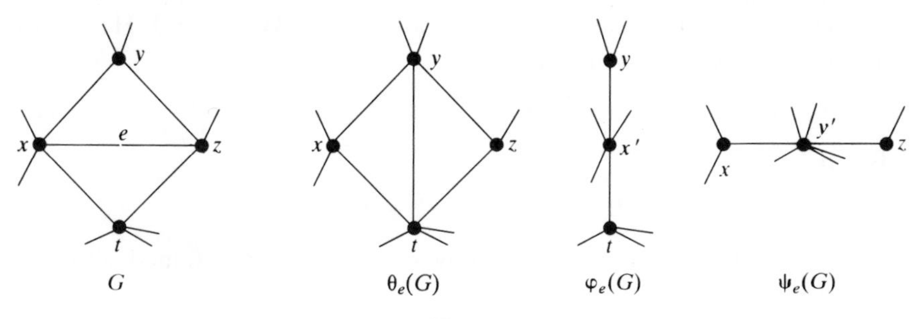

Fig. 5

It is easy to show, using Theorem 2.6, that

$$P(G, \lambda) - P(\theta_e(G), \lambda) = P(\psi_e(G), \lambda) - P(\varphi_e(G), \lambda), \qquad (1)$$

for all λ. The theorem under discussion asserts a somewhat similar linear relation, whose validity is restricted to the case $\lambda = \tau^2$:

Theorem 5.3.

$$P(G, \tau^2) + P(\theta_e(G), \tau^2) = \tau^{-3}\{P(\psi_e(G), \tau^2) + P(\varphi_e(G), \tau^2)\}.$$

Sketch of proof. This theorem can be proved by induction over the number r of vertices and edges of G distinct from e and not belonging to C. If $r = 0$, the verification is trivial. To step from $r = n$ to $r = n + 1$, we have only to use the standard recursion formulas. (For a more detailed proof of this, we refer the reader to Tutte [**28**] or the book of Saaty and Kainen [**26**].) ||

Theorem 5.2 shows that $P(T, \tau^2)$ is quite small when p is fairly large, and so leads us to expect a zero near τ^2, but it does not prove the existence of such a zero. In this connection it is instructive to consider a triangulation T obtained from a tetrahedron by repetition of the operation of subdividing a face into three smaller triangles by new edges drawn to its vertices from a central point. A polyhedron formed in this way is known as a **stack polyhedron**, and the graph formed by its vertices and edges is a special case of the 'chordal graphs' that we shall consider later. Using Theorems 2.4 and 2.5 (the latter with $k = 3$), we find that $P(T, \lambda)$ factorizes completely into linear factors, each equal to λ, $\lambda - 1$, $\lambda - 2$ or $\lambda - 3$. It follows that, for these triangulations, there can be no zero nearer to τ^2 than 2 is, however large p may be.

Theorem 5.2 allows us to assert that $|P(T, \tau^2)|$ is very small for reasonably large p, but Theorem 4.3 tells us that it can never be zero.

We can easily verify Theorem 5.2 for the triangulations of the sphere corresponding to the bipyramids. We note that $\tau^2 - 1 = \tau$, $\tau^2 - 2 = \tau^{-1}$ and $\tau^2 - 3 = -\tau^{-2}$, and that τ^2 is a root of $\lambda^2 - 3\lambda + 1 = 0$. Hence, by Theorem 3.5 with $p = n + 2$,

$$P(B_p, \tau^2) = \tau^2\tau^{-n} + \tau^2 \cdot \tau \cdot (-1)^p\tau^{-2n} = \tau^{-p}\{\tau^4 + (-1)^n\tau^5\tau^{-n}\}.$$

But $n \geq 2$, so that

$$|P(B, \tau^2)| \leq \tau^{-p}\{\tau^4 + \tau^3\} = \tau^{5-p}.$$

The fourth member of the family of theorems mentioned above is called the *Golden Identity*:

Theorem 5.4 (Golden Identity). *Let T be a triangulation having p vertices. Then*

$$P(T, \tau\sqrt{5}) = \sqrt{5} \cdot \tau^{3(k-3)} P^2(T, \tau^2).$$

Sketch of proof. We first note that $\tau\sqrt{5} = \frac{1}{2}(5 + \sqrt{5}) = \tau + 2$.

The proof is by induction on k, combined with a subsidiary induction on the smallest degree (which can only be 2, 3, 4 or 5). The standard recursion formulas enable us to reduce the problem to the case in which G contains no digon. After that, Theorem 5.3 is used, with x in Fig. 5 having the least possible degree. Equation (1) is used for $\lambda = \tau\sqrt{5}$, but both (1) and Theorem 5.3 are available when $\lambda = \tau^2$. Again we refer to the literature (**[30]** or **[26]**) for further details. ||

In combination with Theorem 4.3, Theorem 5.4 gives the interesting result that

$$P(T, \tau\sqrt{5}) > 0.$$

It seems then that $P(T, \lambda)$ is positive at $\lambda = 4$ and at $\lambda = \tau\sqrt{5} = 3.618\ldots$. It is tempting to conjecture that the chromatic polynomial of a triangulation must be positive throughout this interval, but counter-examples are known.

Evidence has been accumulating from specialized corners of the theory that τ^2 is one of an infinite family of numbers having special significance for the chromatic theory of triangulations. The general member of the family is

$$A_n = 2 + 2\cos\{2\pi/n\},$$

n being a positive integer; the numbers A_n are called the **Beraha numbers** by chromocartographers. The first six numbers in this sequence are

$$4, 0, 1, [illegible], \tau + 1 \text{ and } 3,$$

and the chromatic significance of these is indisputable. A tendency for triangulations to have chromatic zeros near A_7 has been noticed, but the effect is much less marked than that for A_5. In the theory of chromatic sums over rooted triangulations, the functional equation for the generating functions used takes a special form when λ has one of the Beraha values (see **[30]**).

The Beraha numbers also arise in other connections. Hall and Lewis **[17]** have studied what they call *constrained* chromatic polynomials. A **constrained chromatic polynomial** gives the number of ways of coloring an n-sided polygon dissected into triangles, subject to certain restrictions on the coloring of the n outer vertices. These restrictions stipulate which vertices must have the same color and which must have different colors. For almost all values of λ, equations can be derived, giving each constrained chromatic polynomial in terms of the ordinary chromatic polynomials of certain related triangulations of the sphere. However, when λ

is the $(n + 1)$th Beraha number, the procedure fails. At least, this is known to be true up to $n = 7$.

Beraha himself drew attention to the fact that $A_n \to 4$ as $n \to \infty$, and suggested that this may explain the very special significance of the number 4 in chromatic map theory.

6. Some Computational Tricks

The computation of specific chromatic polynomials is in many ways more of an art than a science. For graphs of any size, the computation is usually a laborious task, and it is therefore helpful to know some of the 'tricks of the trade', which may serve to make the calculation less burdensome. However, it is not always obvious which tricks, used in which ways, are most likely to speed the computation. We shall now give some methods which may help in finding chromatic polynomials.

The most frequently used method is to apply the Propositions of Recursion given in Theorems 2.5 and 2.6. Note that Theorem 2.6 expresses the chromatic polynomial of a graph in terms of those of the same graph with one edge removed, and another graph obtained from the latter by identifying two vertices, but it can also be used the other way round—that is, in the form

$$P(G - e, \lambda) = P(G, \lambda) + P(G/e, \lambda) \tag{2}$$

—in which case the first graph on the right-hand side is obtained by *adding* an edge to the graph we start with. Thus we have the choice of going in one of two directions: we can remove edges or we can add them. Thus Fig. 6 represents the result of deleting and contracting the edge e from the graph on the left, using Theorem 2.6.

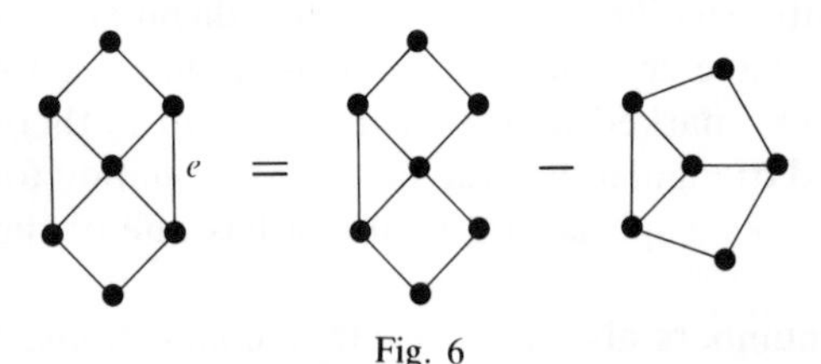

Fig. 6

By applying Theorem 2.6 to a graph, using any suitable edge, then applying it again to the graphs thus formed, and so on, we can eventually express the chromatic polynomial of a graph G in terms of those of edge-free graphs, which we know. If we similarly use equation (2), thus always adding edges, we shall end up expressing $P(G, \lambda)$ in terms of chromatic polynomials of complete graphs, which we also know. In practice, we rarely use either (2) or Theorem 2.6 exclusively, but rather use whichever of these, or other results, will further our ends. Needless to say, if we obtain a graph of any kind whose chromatic polynomial we already know, we need go no further with it, but merely record its contribution to the final result.

Thus the chromatic polynomial of the first graph on the right-hand side of Fig. 6 can be evaluated at once by a double application of Theorem 2.5, as shown in Fig. 7. Its chromatic polynomial is thus expressed in terms of the known chromatic polynomials of circuits.

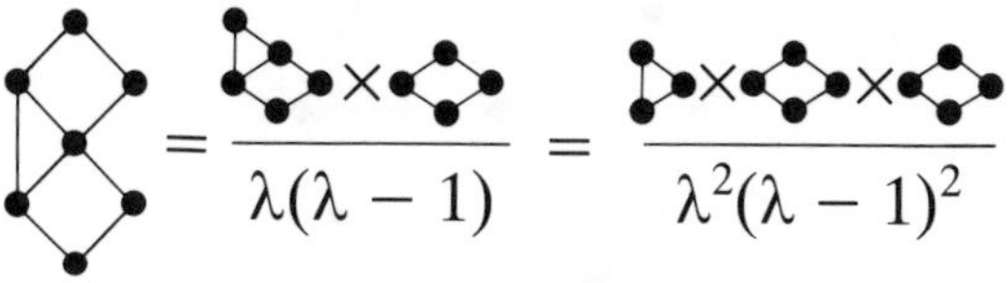

$$= \frac{\quad \times \quad}{\lambda(\lambda - 1)} = \frac{\quad \times \quad \times \quad}{\lambda^2(\lambda - 1)^2}$$

Fig. 7

To the other graph on the right-hand side we apply Theorem 2.6 again, followed by further applications of Theorem 2.5, as shown in Fig. 8.

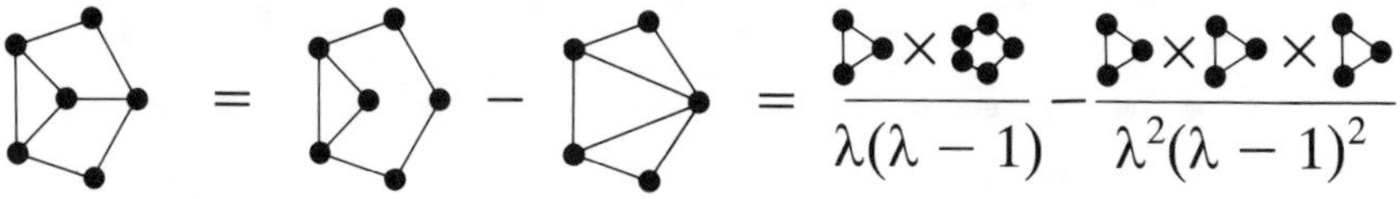

$$= \quad - \quad = \frac{\quad \times \quad}{\lambda(\lambda - 1)} - \frac{\quad \times \quad \times \quad}{\lambda^2(\lambda - 1)^2}$$

Fig. 8

We see that Theorem 2.5 can be extremely useful in that it may express the chromatic polynomial of a graph in terms of those of *much* smaller graphs. In particular, if G has a cut-vertex, then Theorem 2.5 can be applied at once. Similarly, if G is only 2-connected, then we can break G into two graphs using Theorem 2.5—either immediately (if the two vertices in a cutset are adjacent), or after a single application of equation (2). This process is illustrated in Figs. 9 and 10.

This method of using equation (2) can be effectively applied to determine the chromatic polynomials of certain special graphs. A **theta graph**

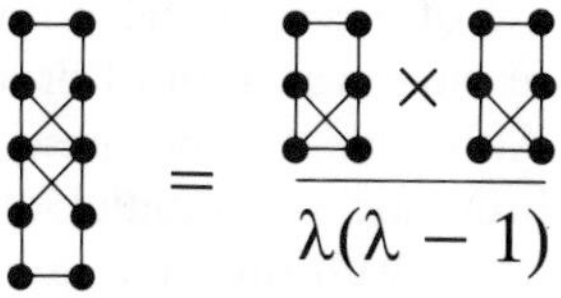

$$= \frac{\quad \times \quad}{\lambda(\lambda - 1)}$$

Fig. 9

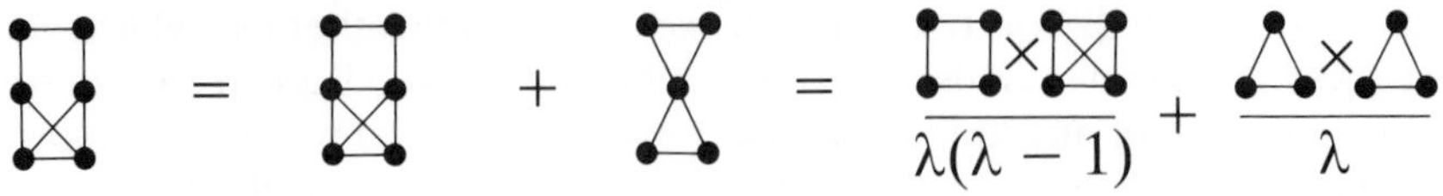

Fig. 10

has two vertices v and w of degree 3, between which are three paths whose other vertices all have degree 2. A typical theta graph is shown in Fig. 11.

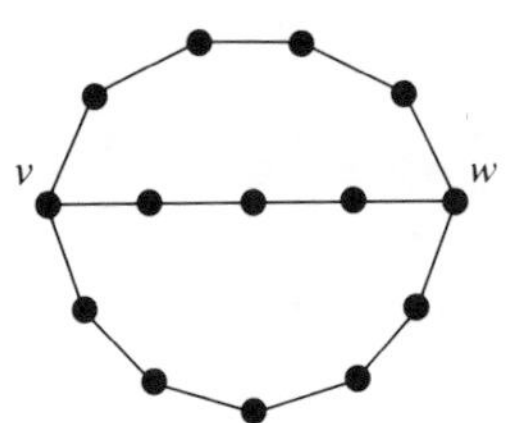

Fig. 11

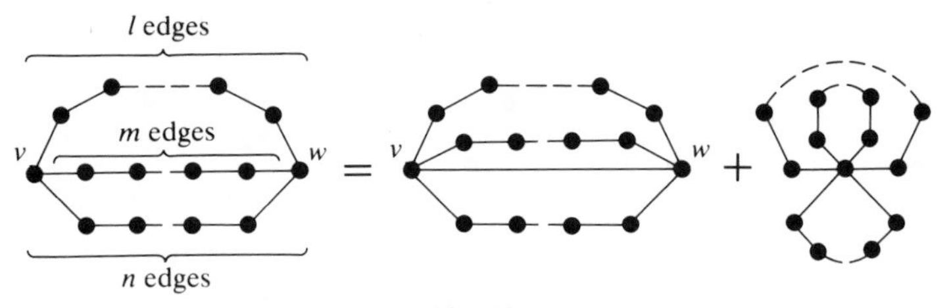

Fig. 12

To find the chromatic polynomial of the theta graph for which the paths have l, m and n edges, we can use (2), adding the edge vw. Diagrammatically this is shown in Fig. 12, and from the graphs on the right of that figure, and by use of Theorem 2.5, we find that the chromatic polynomial of the theta graph is

$$\frac{P(C_{l+1}, \lambda) \cdot P(C_{m+1}, \lambda) \cdot P(C_{n+1}, \lambda)}{\lambda^2(\lambda - 1)^2} + \frac{P(C_l, \lambda) \cdot P(C_m, \lambda) \cdot P(C_n, \lambda)}{\lambda^2}.$$

One of the best tricks of all, when it comes to calculating chromatic polynomials, is to make use of a computer! With the aid of these devices we may well hope to spare ourselves a lot of drudgery, but we should not set our hopes too high. Although it is no great task to write a program that will compute the chromatic polynomial of a given graph, we are likely to find that it will work well only for quite small graphs, and that even for graphs on (say) 10 vertices it may require unconscionable amounts of computer time.

There is a good reason for this. If we know $P(G, \lambda)$ we can easily deduce the chromatic number of G. Hence the calculation of the chromatic polynomial of a graph is at least as difficult as that of finding the chromatic number. This latter problem belongs to the class of *NP-complete problems* (for details, see [**1**]), and it seems highly likely (although nothing certain is yet known) that any algorithm or computer program for solving them must, of necessity, take an amount of time that is an exponential function of the size of the problem. Hence, we must expect a drastic increase in the time taken to compute chromatic polynomials as the number of vertices increases.

There are, of course, various things that we can do to avoid unnecessary computation. When dealing with a graph having many edges, we naturally prefer to use equation (2), taking the route that leads towards complete graphs; when dealing with sparse graphs it is natural to proceed by deleting edges, using Theorem 2.6, thus aiming for edge-free graphs. A computer program of any degree of sophistication will choose one or the other of the alternatives, according to circumstances. Further, it is not necessary to go right to the bitter end, stopping only when we have nothing but complete graphs or edge-free graphs, since we can program the computer to recognize certain kinds of graphs whose chromatic polynomials are known, or easily found.

A program of this type was described by Nijenhuis and Wilf [**24**], which recognizes when it has produced a tree, for which the chromatic polynomial is known. A tree is easily recognized by the fact that $p = q + 1$, provided, of course, that we know that the graph in question is connected. The important feature of the Nijenhuis–Wilf program is that it is organized in such a way that the graphs that it deals with are always connected.

All these tricks, however, are at best palliatives that merely reduce to some extent the labor of calculating chromatic polynomials. They do not alter the basic fact that the computation is an essentially difficult task.

Sometimes it happens that a hand calculation may succeed where a computer program fails. The usual reason for this is that the mind has hit on an *ad hoc* trick, suggested by the graph itself, and this is something that computers are seldom programmed to do. Perceiving and making use of symmetry in a graph is a case in point. Consider the Petersen graph, shown in Fig. 13, which clearly has pentagonal dihedral symmetry. (It actually has more than that, but that is all we shall use.)

Let us apply Theorem 2.6 to each of the edges a, b, c, d and e in turn. The result will be a set of $2^5 = 32$ graphs, each corresponding to a choice

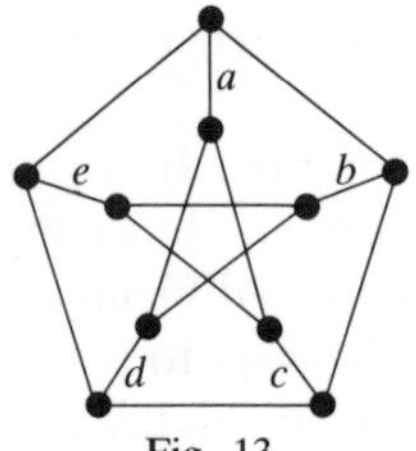

Fig. 13

of five operations, each of which is either deletion or contraction of an edge. Because of the symmetry, many of these graphs will be the same as others. It is easily verified that:

with 5 deletions we get 1 graph, as in Fig. 14(a);
with 4 deletions we get 5 graphs like Fig. 14(b);
with 3 deletions we get 10 graphs like Fig. 14(c);
with 2 deletions we get 10 graphs like Fig. 14(d);
with 1 deletion we get 5 graphs like Fig. 14(e);
with no deletions and all 5 edges being contracted, we get the complete graph on 5 vertices, Fig. 14(f).

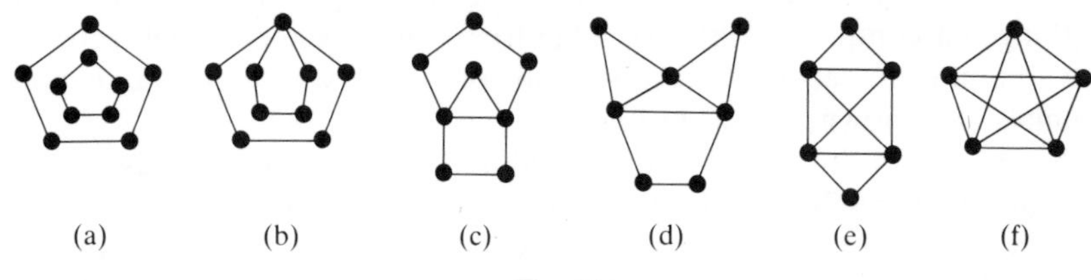

Fig. 14

Now each of these graphs is made up of a number of complete graphs or polygons which have been joined together at a common vertex or along a common edge. Their chromatic polynomials can therefore be found very quickly, by repeated application of Theorem 2.5. These polynomials, with the indicated coefficients and with the correct signs, will then sum to the chromatic polynomial of the Petersen graph. Just as a matter of interest we shall give the answer; it is

$$\lambda(\lambda-1)(\lambda-2)(\lambda^7-12\lambda^6+67\lambda^5-230\lambda^4+529\lambda^3-814\lambda^2+775\lambda-352).$$

Further methods for using the symmetry of a graph to facilitate the computation of its chromatic polynomial are to be found in Chao and Whitehead [**11**].

7. Chromatic Equivalence and Chromatic Uniqueness

As already mentioned, a great deal of the fundamental work on chromatic polynomials was done several decades ago by Whitney, Birkhoff and others (see, for example, **[32]** and **[6]**). Since then, from time to time, new types of problems have arisen to occupy and tax the minds of research workers. A topic that has come to the fore in recent years concerns the related concepts of chromatic equivalence and chromatic uniqueness.

It is very clear that a chromatic polynomial does not generally belong exclusively to one graph. On the contrary, it is possible for a polynomial to be the chromatic polynomial of each of a whole collection (possibly quite large) of graphs. To see an example, we need only look at the trees with p vertices, which all have the same chromatic polynomial—namely, $\lambda(\lambda - 1)^{p-1}$.

Two graphs will be said to be **chromatically equivalent** if they have the same chromatic polynomial. When this happens, it is natural to ask whether there is some reason for it: does it just happen, or is there something about the two graphs that would lead us to expect their chromatic polynomials to be the same? Speculation along these lines gives rise to some interesting questions. Is there some way of recognizing that two graphs are chromatically equivalent, without actually working out the chromatic polynomial for either of them? If we are given a graph, is there a way of constructing other graphs that have the same chromatic polynomial? And so on.

For some kinds of graph, answers to these questions are at hand. Consider, for example, two graphs that consist of the same two blocks, but put together at a cut-vertex in different ways, as indicated in Fig. 15.

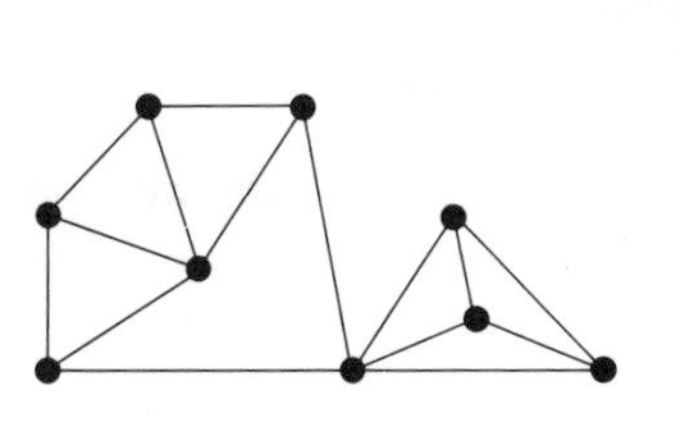

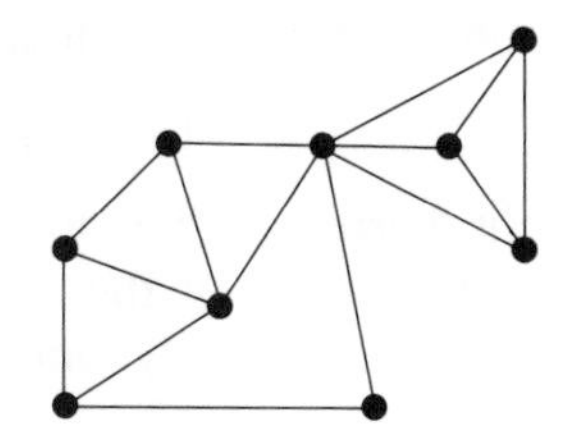

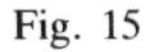

Fig. 15

By Theorem 2.5, we see that the chromatic polynomial in each case is the product of the chromatic polynomials of the blocks divided by λ. It follows immediately that these two graphs are chromatically equivalent. Moreover, since Theorem 2.5 applies to graphs that overlap in any complete graph (not just in a single vertex), it can be applied in a more general setting. Thus, if we take two graphs G and H and identify one edge of G with one of H, we obtain a graph whose chromatic polynomial can be found from Theorem 2.5; if we identify a different pair of edges (one from G, one from H) we get, in general, a different graph, but the chromatic polynomial will be the same.

There is an interesting family of graphs that can be built up in this sort of way, and whose chromatic polynomials assume a particularly simple form. These are the *chordal graphs*, originally introduced by Dirac [**13**] under the name of *rigid circuit graphs*. A **chordal graph** is one in which every circuit of length greater than 3 has a chord. It can be shown (see, for example, Gavril [**15**]) that a graph is a chordal graph if and only if it can be built up step by step, starting with a single vertex, by repeated application of the following construction:

introduce a new vertex, and join it to any set of mutually adjacent vertices in the present graph.

The general form of the chromatic polynomial of a chordal graph follows immediately from this constructive definition. If the new vertex is joined to r vertices in the graph, then since these vertices (being mutually adjacent) require r colors, there are $\lambda - r$ colors available for the new vertex. Hence each step of the construction contributes a factor $\lambda - r$ to the chromatic polynomial, which must therefore be of the form

$$\lambda^{m_0}(\lambda - 1)^{m_1}(\lambda - 2)^{m_2} \dots (\lambda - k)^{m_k}, \tag{3}$$

for some integer k, where $m_0, m_1, \dots, m_k$ are positive integers. Clearly, $k + 1$ is the size of the largest complete subgraph (clique) in the graph. Conversely, given a set of such numbers we can usually construct several graphs having (3) as chromatic polynomial. Figure 16 shows two of the many graphs that have chromatic polynomial

$$\lambda(\lambda - 1)^2(\lambda - 2)^3(\lambda - 3)^2.$$

The graphs corresponding to the stack polyhedra that we considered earlier form a subset of the set of chordal graphs, and as we observed then, their chromatic polynomials are of this general form.

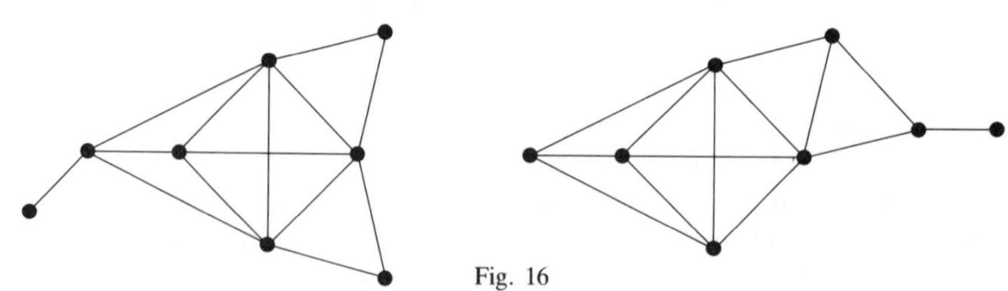

Fig. 16

It is tempting to think that any graph whose polynomial is of the form (3) is a chordal graph, but this is not so. The graph of Fig. 17 is not chordal, but its chromatic polynomial is

$$\lambda(\lambda - 1)(\lambda - 2)(\lambda - 3)^3(\lambda - 4),$$

as one can easily verify.

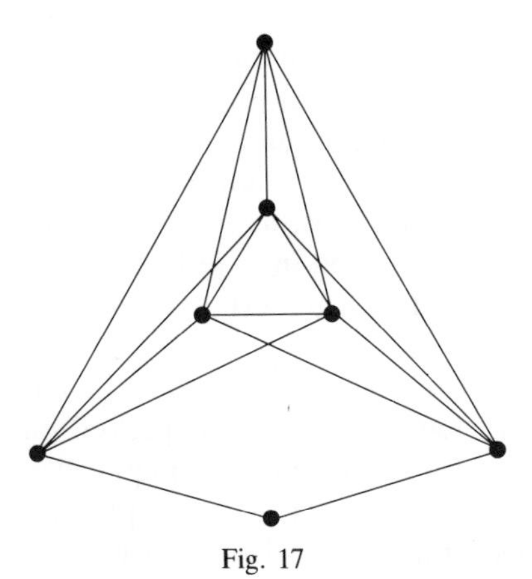

Fig. 17

There is an interesting construction for chromatically equivalent graphs that uses *rotors*. A **rotor** in a graph G is a subgraph R possessing a rotational symmetry θ, such that all the vertices of attachment of R to the rest of the graph are equivalent under θ. The **order** of the rotor is the order of its automorphism θ. Commonly, we can change G into a different graph G' by 'flipping' the rotor R—that is, by replacing it by its mirror image.

An interesting general question arises: what properties of G are invariant under the operation of rotor-flipping? It is well known that one such property is the number of spanning trees (see [**8**]). It can be shown that the chromatic polynomial of G is another such property, provided that the order of the rotor concerned does not exceed 5; for details, see Tutte [**31**]. By exploiting this fact, it is possible to get chromatically equivalent graphs, even planar ones, of connectivity 5.

There are non-planar examples, due to Foldes [**14**], in which the flipping of a rotor of order 6 or more changes the chromatic polynomial. It has been shown by Lee [**19**] that the flipping of a rotor of order 6 in a planar triangulation cannot change the chromatic polynomial. So far, no case is known in which the flipping of a rotor in a planar graph G does change the chromatic polynomial.

Passing now to the other extreme, we look at those graphs which have a chromatic polynomial all to themselves—that is, graphs G and H such that

$$P(H, \lambda) = P(G, \lambda)$$

implies that H is isomorphic to G. Such a graph is said to be **chromatically unique**.

Some trivial examples come immediately to mind. The complete graphs and the edge-free graphs are clearly in this category, and so is the only tree on three vertices. Further examples can be culled from catalogues of graphs on a fixed number of vertices, by working out all the chromatic polynomials (preferably using a computer!). Some information on chromatically unique graphs has been obtained in this way (although only for graphs with at most nine vertices, as far as we know), but what is much more challenging is to try to find infinite families of chromatically unique graphs. This is quite a difficult undertaking, and not many such families are known.

In [**12**], Chao and Whitehead showed that the circuit graphs C_p are all chromatically unique. Later, Loerinc [**21**] proved the more general result that all theta graphs are chromatically unique. The same authors have produced several other such families of graphs. Some likely candidates have proved disappointing. The wheels up to W_5 are all chromatically unique, and so is W_7; but surprisingly W_6 is not, since Chao and Whitehead [**12**] noted that it shares its chromatic polynomial with the graph of Fig. 18. A computer investigation by one of us a few years ago revealed that W_8 is not chromatically unique, since it has the same chromatic polynomial as 10 other graphs. This investigation has been extended to show that W_9 and W_{10} are both chromatically unique.

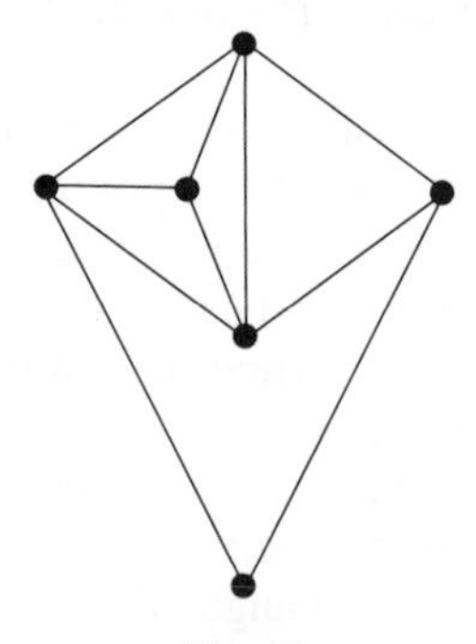

Fig. 18

The study of chromatic equivalence and chromatic uniqueness is a promising area for research—one in which the questions are easy to put, but tantalizingly difficult to answer.

8. Unsolved Problems

There are many questions relating to chromatic polynomials to which no answers are known. Some of them have been around for a very long time. In this section we present a few.

Characterization of Chromatic Polynomials

Is there a necessary and sufficient condition for a polynomial to be the chromatic polynomial of some graph? In other words, is there some simple test that will enable us to determine whether there is a graph whose chromatic polynomial is a given polynomial?

We have seen already that there are several known *necessary* conditions—the polynomial must be monic, have terms that alternate in sign, have no constant term and so on—but these conditions, even taken all together, are not sufficient. If we ignore the word "simple" above, we can easily describe a method of testing whether a polynomial is chromatic. From the polynomial we know the number of vertices and edges the graph must have (if it exists at all), for the degree of the polynomial gives us the number p of vertices, and the number of edges is given by the coefficient of λ^{p-1}. We then, very laboriously, generate all such graphs and find their chromatic polynomials. This would work, but it is clearly not a practical test, and certainly not simple in any sense of the word.

Chromatic Uniqueness and Chromatic Equivalence

We can ask for a method of recognizing whether a given graph is chromatically unique, but it is unlikely that any easy answer will be forthcoming. For graphs on up to 10 vertices, the existence of a catalogue of all such graphs (see Cameron *et al.* [**10**]) provides a routine, although tedious, means of resolving the question. But for larger graphs it is extremely difficult to prove chromatic uniqueness. The results that are known were obtained by painstaking detective work—by carefully wringing from a chromatic polynomial every drop of information about the graph, until it could be shown that there was only the one possibility. As things stand at present, there seems to be no other way of approaching this problem.

Much the same can be said about chromatic equivalence. It is, of course, straightforward to test whether two graphs are chromatically equivalent, but to determine whether there are other graphs with the same chromatic polynomial is quite another problem. We saw in Section 7 how we can take some graphs apart at cut-vertices and cut-sets, and reassemble them to get a different graph with the same chromatic polynomial. This prompts speculation on the possibility of an algorithm which will generate *all* the graphs chromatically equivalent to a given graph, or (what amounts to the same thing) generate all the graphs with a given chromatic polynomial $P(\lambda)$.

Given the present state of the art this seems altogether too much to hope for. It might seem less sanguine to ask for an algorithm that would merely compute the *number* of graphs whose chromatic polynomial is $P(\lambda)$, but this is probably just as intractable as the problem of generating them. This last problem is clearly an extension of the first problem in this section, that of characterizing chromatic polynomials.

The Unimodal Conjecture

All the above problems seem to be hopelessly difficult—very little progress has been made with them apart from a few partial results. A conjecture about which one can perhaps feel somewhat more hopeful is the *unimodal conjecture* for the coefficients of chromatic polynomials, a conjecture that was mentioned at the end of Section 2. There it was stated in the form:

the statement

$$a_j > a_{j+1} \text{ and } a_{j+1} < a_{j+2} \tag{1}$$

is false for all j.

This is the conjecture originally given in **[24]**, but it seems likely that it can, and should, be replaced by a conjecture of Hoggar **[18]**—namely, that

$$a_j a_{j+2} < {a_{j+1}}^2 \text{ holds for all } j. \tag{2}$$

This is a stronger assertion, since it implies (1) but is not implied by it; it also seems more 'natural', in that the property defined by (2)—known as *strong logarithmic concavity*—is preserved under multiplication of polynomials, whereas property (1) may not be. As we have already seen, the property of being a chromatic polynomial is also preserved under multiplication—the product of two chromatic polynomials is a chromatic polynomial.

The coefficients that arise when a chromatic polynomial is expressed in terms of those of complete graphs also seem to exhibit a unimodal distribution, and it has been conjectured that they too have the property of strong logarithmic concavity. Yet another sequence of coefficients arises when a chromatic polynomial is expressed in terms of those of trees, as with the Nijenhuis–Wilf algorithm. This is equivalent to writing $P(G, \lambda)/\lambda$ with $\lambda - 1$ as the variable, instead of λ. For these coefficients the situation in rather different; the unimodal property does not hold for many disconnected graphs, and examples are known (see **[22]**) of connected graphs for which three or more consecutive coefficients are equal, thus contradicting the property of *strong* logarithmic concavity. However, no counter-examples are known to the slightly weaker conjecture that these coefficients have the property of (weak) logarithmic concavity, defined by replacing $<$ in (2) by $\leq$.

Hence, for each of the three ways of expressing the chromatic polynomial, we have a unimodal conjecture to the effect that the coefficients have the property of logarithmic concavity. For the expression in terms of trees we have to restrict the conjecture to connected graphs, while for the other two we can sharpen the conjecture to assert strong logarithmic concavity, since for them no examples are known of three consecutive equal coefficients. Exhaustive examination has recently shown that the three conjectures hold for all graphs on less than 10 vertices.

Chromatic Polynomials of Families of Graphs

A more tractable genre of research problem is the investigation of the chromatic polynomials of certain infinite families of graphs. We have already seen some examples of this in the general formulas given for the chromatic polynomials of trees (Theorem 3.1), of circuits (Theorem 3.2), and of wheels and bipyramids (Theorem 3.3). These results were fairly easy to obtain, but it frequently requires considerable ingenuity to derive results of this kind.

It is well known that every graph M on a sphere has a dual graph M^*. The dual of the bipyramid B of Fig. 4 is a graph that can be variously described as a prism (or, more strictly, the edge-skeleton of a prism), as a sort of squirrel-cage or as an endless ladder. It is shown in Fig. 19. A

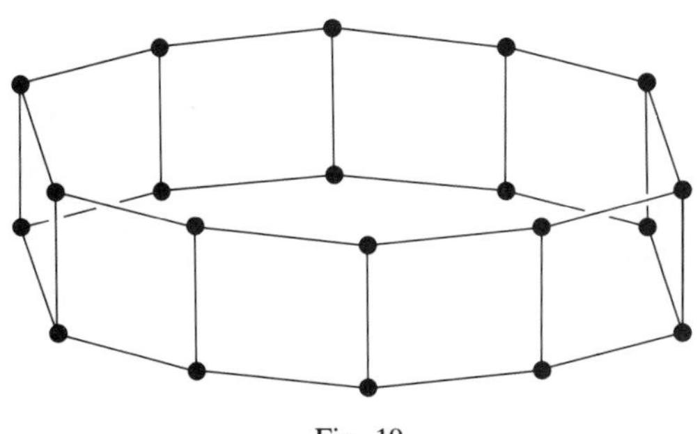

Fig. 19

general formula for the chromatic polynomials of these graphs (now usually referred to as *ladders*) has been given by Biggs *et al.* [**4**].

Unfortunately, there appears to be no connection between the chromatic polynomial of a plane graph and that of its dual, and compared with the simple form of Theorem 4.6 the chromatic polynomials for ladders are considerably more complicated.

One can also define *Möbius ladders*—non-planar graphs obtained from ladders by crossing the 'sides' of a ladder between two consecutive 'rungs'. In this way, the sides now form the boundary of a Möbius strip (Fig. 20). The chromatic polynomials of these graphs are also found in [**4**].

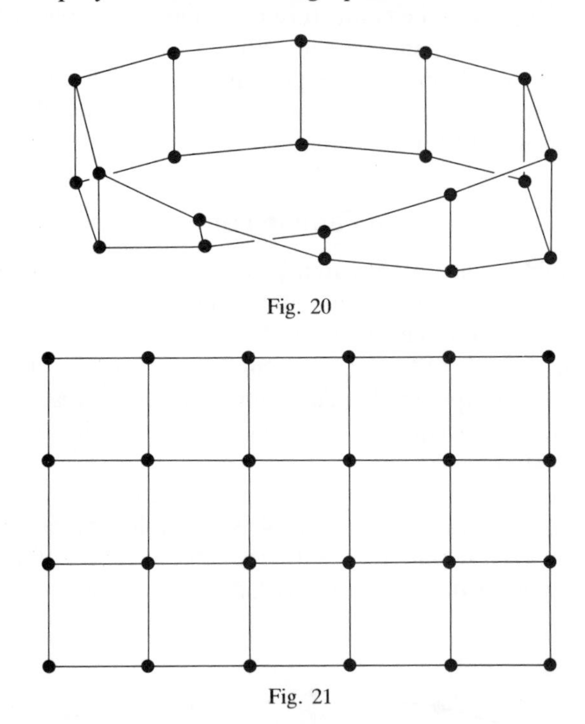

Fig. 20

Fig. 21

Again, there appears to be no connection between the chromatic polynomials of a graph and its complement. Hence the complements of any family of graphs can form a separate topic for investigation. Some interesting work has been done by Chao and Whitehead [**12**] and by Loerinc [**22**] on the chromatic polynomials of the complements of trees and of theta graphs.

Even within this area of research—that of finding general formulas for chromatic polynomials of families of graphs—it is remarkably easy to formulate exceedingly difficult problems. Here is an example:

Problem 8.1. *Find a general formula for the chromatic polynomial* $P(X_{m,n}, \lambda)$ *of the graph* $X_{m,n}$ *associated with an* $m \times n$ *'chessboard'—the graph with vertices at the integer lattice points* (x, y) *with* $1 \le x \le n$, $1 \le y \le m$, *and in which each vertex is joined to the four or fewer vertices at distance* 1 *from it.* (Figure 21 shows the graph $X_{4,6}$.)

This is an easy question to ask, but without doubt a fiendishly difficult one to answer. It does not seem to be altogether unreasonable, however, and it is perhaps within the bounds of possibility that such a formula could be found.

References

1. A. V. Aho, J. E. Hopcroft and J. D. Ullman, *The Design and Analysis of Computer Algorithms*, Addison-Wesley, Reading, Mass., 1974; *MR*54#1706.
2. K. Appel and W. Haken, Every planar map is four colorable: Part 1, Discharging, *Illinois J. Math.* **21** (1977), 429–490; *MR*58#27598a.
3. K. Appel, W. Haken and J. Koch, Every planar map is four colorable: Part 2, Reducibility, *Illinois J. Math.* **21** (1977), 491–567; *MR*58#27598b.
4. N. L. Biggs, R. M. Damerell and D. A. Sands, Recursive families of graphs, *J. Combinatorial Theory (B)* **12** (1972), 123–131; *MR*45#3245.
5. G. D. Birkhoff, A determinantal formula for the number of ways of coloring a map, *Ann. of Math.* **14** (1912), 42–46.
6. G. D. Birkhoff, On the polynomial expressions for the number of ways of coloring a map, *Ann. Scuola Norm. Sup. (Pisa) (2)* **3** (1934), 85–104, and *Collected Mathematical Papers, Vol. 3*, Amer. Math. Soc., New York, 1950, pp. 29–47.
7. G. D. Birkhoff and D. C. Lewis, Chromatic polynomials, *Trans. Amer. Math. Soc.* **60** (1946), 355–451; *MR*8–284f.
8. R. L. Brooks, C. A. B. Smith, A. H. Stone and W. T. Tutte, The dissection of rectangles into squares, *Duke Math. J.* **7** (1940), 312–340; *MR*2–153d.
9. L. Caccetta and S. Foldes, Symmetric calculation of chromatic polynomials, *Ars Combinatoria* **4** (1977), 289–292.
10. R. D. Cameron, C. J. Colbourn, R. C. Read and N. C. Wormald, On the construction of a catalogue of all graphs on 10 vertices, *J. Graph Theory* **9** (1985), 551–562.
11. C. Y. Chao and E. G. Whitehead, On chromatic equivalence of graphs, *Theory and Applications of Graphs* (ed. Y. Alavi and D. R. Lick), Lecture Notes in Math. **642**, Springer, Berlin, 1978, pp. 121–131; *MR*58#21753.
12. C. Y. Chao and E. G. Whitehead, Chromatically unique graphs, *Discrete Math.* **27** (1979), 171–177; *MR*80m:05036.
13. G. A. Dirac, On rigid circuit graphs, *Abh. Math. Sem. Univ. Hamburg* **25** (1961), 71–76; *MR*24#A57.
14. S. Foldes, The rotor effect can alter the chromatic polynomial, *J. Combinatorial Theory (B)* **25** (1978), 237–239; *MR*80c:05070.
15. F. Gavril, The intersection graphs of subtrees in trees are exactly the chordal graphs, *J. Combinatorial Theory (B)* **16** (1974), 47–56; *MR*48#10868.
16. D. Gernert, Recent results on chromatic polynomials, *Methods Oper. Res.* **51** (1984), 307–314; *MR*86g:05033.
17. D. W. Hall and D. C. Lewis, Coloring six-rings, *Trans. Amer. Math. Soc.* **64** (1948), 184–191; *MR*10–136g.
18. S. Hoggar, Chromatic polynomials and logarithmic concavity, *J. Combinatorial Theory (B)* **16** (1974), 248–254; *MR*49#7170.

19. L. A. Lee, On chromatically equivalent graphs, Ph.D. Thesis, The George Washington University, 1975.
20. W. X. Li and F. Tian, Some notes on the chromatic polynomials of graphs (Chinese), *Acta Math. Sinica* **21** (1978), 223–230.
21. B. Loerinc, Chromatic uniqueness of the generalized θ-graph, *Discrete Math.* **23** (1978), 313–316; *MR***80a**:05095.
22. B. Loerinc, Computing chromatic polynomials for special families of graphs, Ph.D. Thesis, New York University, 1979.
23. G. H. J. Meredith, Coefficients of chromatic polynomials, *J. Combinatorial Theory (B)* **13** (1972), 14–17; *MR***46**#8886.
24. A Nijenhuis and H. S. Wilf, *Combinatorial Algorithms. For Computers and Calculators*, Academic Press, New York, 1978; *MR***80a**:68076.
25. R. C. Read, An introduction to chromatic polynomials, *J. Combinatorial Theory* **4** (1968), 52–71; *MR***37**#104.
26. T. L. Saaty and P. C. Kainen, *The Four-color Problem. Assaults and Conquest*, McGraw-Hill, New York, 1977; *MR***58**#246.
27. V. Thier, Graphen and Polynome, Diploma Thesis, TU München, 1983.
28. W. T. Tutte, On chromatic polynomials and the golden ratio, *J. Combinatorial Theory* **9** (1970), 289–296; *MR***42**#7557.
29. W. T. Tutte, More about chromatic polynomials and the golden ratio, *Combinatorial Structures and their Applications* (ed. R. K. Guy *et al.*), Gordon and Breach, New York, 1970, pp. 439–453; *MR***41**#8299.
30. W. T. Tutte, Chromatic sums for planar triangulations, V: Special equations, *Canad. J. Math.* **26** (1974), 893–907; *MR***50**#167.
31. W. T. Tutte, Codichromatic graphs, *J. Combinatorial Theory (B)* **16** (1974), 168–174; *MR***48**#10876.
32. H. Whitney, A logical expansion in mathematics, *Bull. Amer. Math. Soc.* **38** (1932), 572–579.
33. H. Whitney, On the coloring of graphs, *Ann. of Math.* **33** (1932), 688–718.
34. H. S. Wilf, Which polynomials are chromatic?, *Colloq. Internaz. sulle Teorie Combinatorie, 1973* (ed. B. Segre), *Atti dei Convegni Lincei 17*, Rome, 1976, pp. 247–256; *MR***56**#11841.
35. D. R. Woodall, Zeros of chromatic polynomials, *Combinatorial Surveys: Proc. Sixth British Combinatorial Conference* (ed. P. J. Cameron), Academic Press, London, 1977, pp. 199–223; *MR***57**#2974.
36. A. A. Zykov, On some properties of linear complexes, *Amer. Math. Soc. Trans.* **79** (1952), translated from *Mat. Sbornik N.S.* **24(66)** (1949), 163–188; *MR***11**–733h and **14**–493a.

3
Matroids and their Applications

DOMINIC WELSH

1. Introduction

Just as point set topology is an abstraction of the open sets of Euclidean space without the use of distance, so H. Whitney's pioneering paper [**50**] on matroids in 1935 was an attempt to axiomatize the concept of linear independence without reference to vectors, and in so doing to unify concepts in graph theory and linear algebra.

Apart from isolated papers by Birkhoff [**4**], MacLane [**21**], [**22**], [**23**], and Rado [**29**], [**30**], Whitney's paper seems to have been ignored until the late 1950s when W. T. Tutte published a series of crucial papers [**37**], [**38**], characterizing (among other things) those matroids which arise from graphs.

In 1965 it was realized that, apart from the graph-theoretic and vector space matroids known to Whitney in 1935, matroids arise very naturally and are of enormous help in simplifying and unifying a host of results in that area of combinatorics known as *transversal theory*. A comprehensive account of this development may be found in the book by Mirsky [**24**], where matroids (not restricted to finite sets) are called *independence structures*.

In a parallel development, a suggestion by Birkhoff [**4**] that matroid theory was just a particular case of the theory of semimodular lattices was

GRAPH THEORY, 3
ISBN 0-12-086203-4

followed up by Rota [**31**]; for an account of this approach, see the influential monograph of Crapo and Rota [**9**]. An account of the general theory and its applications can now be found in several recent texts on combinatorics and optimization theory—for example, Aigner [**1**], Berge [**3**], Bryant and Perfect [**6**], Brylawski and Kelly [**8**], Lawler [**20**], Tutte [**43**], Welsh [**47**] and Wilson [**51**].

This article is roughly divided into two parts. In the first part we survey the bare bones of the subject as outlined above, paying particular attention to those parts of greatest appeal to a reader whose main interest is in graph theory. The second part differs in that it attempts to place the reader at the frontier of the most interesting present-day research problems. To do this we have to be selective, and we have chosen to treat in a little more depth some remarkable advances made in the last few years, notably by P. D. Seymour. One of the attractions of matroid theory is that it usually provides several different natural settings in which to view a problem, and the theme of this second part is the use of matroids to relate problems about colorings and flows in graphs with problems of projective geometry.

In 1912, O. Veblen [**44**] was the first to attempt to settle the four-color conjecture by geometrical methods. Following on from this, Tutte [**41**] in 1966 formulated a fascinating geometrical conjecture—his *tangential block conjecture*, asserting that there are just three minimal geometrical configurations which have a non-empty intersection with each flat of co-dimension 2 in the projective geometry $PG(n,2)$. (He called them *tangential 2-blocks*.) As a result of his work on chromatic polynomials, Tutte had earlier made a number of equally interesting conjectures about flows in networks (see [**36**]). One of these was proved in 1975 by F. Jaeger [**18**], who gave an elegant proof that every bridgeless graph has an 8-flow—or equivalently, that every bridgeless graph is the union of three Eulerian subgraphs. More recently, a remarkable advance towards the solution of Tutte's tangential block conjecture was made by Seymour [**34**], who showed that the only new tangential blocks must be cocycle matroids of graphs. This reduced a seemingly intractable geometrical question to the conceptually easier problem of classifying those graphs whose edge-sets cannot be covered by two Eulerian subgraphs. This area of combinatorics fits very easily into matroid theory, since the latter provides a natural framework from which it is convenient to use either geometric or graph-theoretic arguments. The key concept is the idea of extending the chromatic polynomial of a graph to arbitrary matroids. For simple matroids this becomes the characteristic polynomial of the corresponding geometric lattice, and is one of the nicest examples of the Tutte–Grothendieck theory developed by T. H. Brylawski [**7**] in 1972.

Finally, in Section 8, we highlight what is probably the most important theorem so far developed by matroid theory. This is Seymour's synthesis of totally unimodular matrices from graphs and one other special 5×10 matrix. First, however, in Sections 2–6 we cover what are now regarded as the fundamentals of the subject.

2. Matroid Definitions

In this section we define the basic concepts of matroid theory and show how matroids can be defined axiomatically in terms of quite different concepts.

First we introduce some notation: $V(r, q)$ denotes the vector space of rank r (and hence dimension $r - 1$) over the field $GF(q)$; $PG(r - 1, q)$ denotes the corresponding projective space. We shall move freely between vector spaces and the corresponding projective spaces. Much of the time we shall work with the field $GF(2)$, where the difference between the geometry and the corresponding vector space is minimal.

For the most part, our graph terminology is standard. In order to avoid confusion with the corresponding matroid concept, we define a **cycle** of a graph G to be a set of edges $\{e_1, \ldots, e_t\}$ such that e_i and e_{i+1} are adjacent for $1 \leq i < t$ and e_i is not adjacent to e_j for $j \neq i + 1$ or $i - 1$, except that e_t is adjacent to e_1. In other words, a cycle is the set of edges of a simple closed walk. The deletion of an edge e from a graph G gives a graph G'_e on the same vertex-set. The contraction of e from G gives a graph G''_e which is obtained from G by deleting e and then identifying its two ends. If G and H are two graphs, then G is said to be **contractible** to H, or to have H as a **minor**, if we can obtain H from G by an appropriate sequence of deletions and contractions of edges. A **cutset** of G is a set of edges whose removal increases the number of connected components of G, and a **cocycle** is a minimal cutset; a **bridge** or **isthmus** is a cutset of size 1.

Consider the following two situations:

Example 2.1. Let S be a finite collection of vectors from a vector space. If X and Y are linearly independent sets of vectors of S with $|X| = k$ and $|Y| = k + 1$, then there exists $y \in Y \backslash X$ such that $X \cup \{y\}$ contains $k + 1$ linearly independent vectors.

Example 2.2. Let G be a graph. If X and Y are subsets of edges with $|Y| = |X| + 1$ and such that the subgraphs induced by X and Y are both acyclic, then there exists an edge $y \in Y \backslash X$ such that $X \cup \{y\}$ is also acyclic.

The 'exchange property' common to the above examples is what makes sets of vectors in vector spaces and edges in graphs into examples of matroids.

A matroid can be defined in several ways, using different axiom systems in terms of different concepts. We choose to start with the definition involving 'independent sets'.

Independence Axioms. *A* **matroid** *M is a finite set S and a family $\mathcal{I}$ of subsets of S, called* **independent sets**, *such that*:

($I1$) $\emptyset \in \mathcal{I}$;

($I2$) *if $A \in \mathcal{I}$, then every subset of A is a member of $\mathcal{I}$*;

($I3$) *if $X, Y \in \mathcal{I}$ and $|Y| = |X| + 1$, then there exists $y \in Y \backslash X$ such that $X \cup \{y\} \in \mathcal{I}$.*

For any subset A of S, the **rank of A**, denoted by $r(A)$, is the cardinality of a maximal independent subset of A. It is easy to verify that the rank function r is well defined and satisfies the *submodular inequality*

$$r(A \cup B) + r(A \cap B) \leq r(A) + r(B),$$

for any $A, B \subseteq S$. The **rank of M**, written $r(M)$, is the rank of the set S in M.

A **base** of M is any maximal independent subset of S. Since the rank function is well defined, all bases of M have the same cardinality—namely, the rank $r(M)$ of the matroid.

It is clear that a knowledge of the collection of bases of a matroid uniquely determines the matroid; a set is 'independent' if and only if it is a subset of a base. Indeed we could have started with bases since we can define a matroid in terms of them, as follows:

Base Axioms. *A* **matroid** *M is a finite set S and a family $\mathcal{B}$ of subsets of S, called* **bases**, *such that*:

($B1$) *if B and B' are distinct members of $\mathcal{B}$, then $B \nsubseteq B'$*;

($B2$) *if B, B' are members of $\mathcal{B}$, and if $x \in \mathcal{B}$, then there exists $y \in B'$ such that $(B - \{x\}) \cup \{y\}$ is also a member of $\mathcal{B}$.*

The equivalence of these axioms with the independence axioms is proved in Whitney's original paper. It is now easy to define the most familiar class of matroids arising from graphs:

Theorem 2.1. *Associated with any graph G is a matroid $M(G)$, defined on the edge-set $E(G)$, whose independent sets are the acyclic subgraphs of G. The bases of $M(G)$ are the maximal spanning forests of G and $r(M(G)) = |V(G)| - k(G)$, where $k(G)$ is the number of connected components of G.* $\|$

The matroid $M(G)$ is known as the **cycle matroid** of G.

Two matroids M and M' on sets S and S' are called **isomorphic** if there is a one-to-one map f from S to S' which preserves independence. Thus, if G and H are isomorphic graphs, then $M(G)$ and $M(H)$ are isomorphic matroids. However, the cycle matroids of two non-isomorphic graphs may be isomorphic, as illustrated in Fig. 1. Indeed, if G and H are any two forests with the same number of edges, then their cycle matroids are isomorphic.

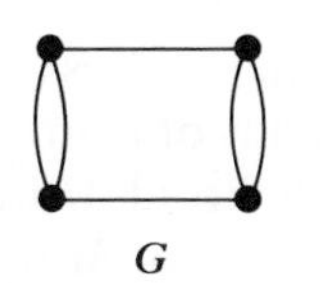

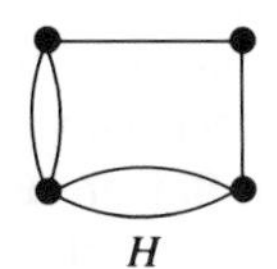

Fig. 1

Two matroids M_1 and M_2 on disjoint sets S_1 and S_2 can be combined to form the **direct sum** $M_1 + M_2$ on $S_1 \cup S_2$, whose bases are all sets of the form $B_1 \cup B_2$ where B_i is a base of M_i. When G_1 and G_2 are disjoint graphs, $M(G_1) + M(G_2)$ is just the cycle matroid of $G_1 \cup G_2$.

A subset X of S is **dependent** in a matroid M if it is not independent; thus a set of edges of a graph G is dependent in $M(G)$ if and only if it contains a cycle. Accordingly, we call X a **circuit** of M if it is a minimal dependent set. Similarly, we say that two elements x and y of S are **parallel** in M if $\{x, y\}$ is a circuit of M, and x is a **loop** of M if $\{x\}$ is a circuit of M. These terms have an obvious graphical origin.

It is obvious that a matroid is also determined by its circuits. In his original paper Whitney gave axioms for a matroid in terms of its circuits. One such system is the following:

Circuit Axioms. *A* **matroid** *M is a finite set S and a family $\mathscr{C}$ of subsets of S, called* **circuits**, *such that*:

(C1) *no member of $\mathscr{C}$ properly contains another*;

(C2) *if C_1 and C_2 are members of $\mathscr{C}$ and $x \in C_1 \cap C_2$, then $(C_1 \cup C_2)\backslash\{x\}$ contains a member of $\mathscr{C}$.*

Another axiom system for matroids is the following:

Closure Axioms. *A* **matroid** *M is a finite set S and a* **closure operator** *σ on the subsets of S, such that*:

(S1) *if $A \subseteq B$, then $\sigma(A) \subseteq \sigma(B)$*;

(S2) *if $A \subseteq S$, then $A \subseteq \sigma(A) = \sigma\sigma(A)$*;

(S3) *if $x \in \sigma(A \cup \{y\})$ and $x \notin \sigma(A)$, then $y \in \sigma(A \cup \{x\})$.*

This last axiom is sometimes called the *Steinitz–MacLane exchange axiom*. It is easy to see that an element x belongs to $\sigma(A)$ if and only if $r(A \cup \{x\}) = r(A)$. From this it is clear that, for a graph G, the closure of any set A of edges consists of A together with all edges x lying in some cycle C contained in $A \cup \{x\}$.

The four axiom systems stated above are mutually equivalent, and are just a few of the many different systems appearing in the literature. For the most part the proofs of equivalence are routine, although sometimes laborious; see, for example, Whitney **[50]** or Welsh **[47]**.

There are also matroid concepts which have a less obvious graphical origin but which reflect the vector space origins of matroids. For example, a subset F of S is a **flat** or **closed set** of a matroid M if, for each $y \notin F$, $r(F \cup \{y\}) > r(F)$—in other words, F is a flat of M if and only if $\sigma F = F$. A **hyperplane** is a maximal proper flat—that is, if M has rank r then a hyperplane is any flat of rank $r - 1$.

Example 2.3 (Uniform matroids). Let S be a set of n elements, and let $U_{k,n}$ be the matroid on S whose bases are those subsets of S with exactly k elements. The independent sets of $U_{k,n}$ are those subsets of S with k or fewer elements, the circuits are those subsets with exactly $k + 1$ elements, and the hyperplanes are those subsets with $k - 1$ elements. We call $U_{k,n}$ a **uniform matroid**.

In general, the flats of a matroid, ordered by inclusion, form a geometric lattice, and every geometric lattice can be obtained from a matroid in this way. In fact, there is a natural many-to-one correspondence between matroids and geometric lattices, in which the loops of the matroid are identified with the zero element of the lattice, and sets of mutually parallel elements are identified with the *atoms* (elements of rank one), in exactly the same way as projective spaces are obtained from vector spaces by identifying elements which are mutually linearly dependent.

This intimate relationship between matroids and geometric lattices was the main observation of Birkhoff's paper **[4]** in 1935. Although useful, in that it places matroid theory in perspective as essentially a branch of geometry, it does not seem (at the moment at least) to be a very convenient context in which to treat one of the most useful combinatorial aspects of matroid theory—namely, matroid duality.

3. Duality and Minors

Duality is one of the most important concepts in matroid theory. It corresponds to orthogonality in the theory of vector spaces, but is really much simpler. If M is a matroid on S, its **dual matroid** M^* is the matroid on S whose bases are the complements of the bases of M. Clearly $(M^*)^* = M$.

Given an element e of S, there are two useful operations on M which involve e. We first let M'_e be the matroid on $S \backslash e$ whose independent sets are those independent sets of M which are contained in $S \backslash e$; we call this operation the **deletion of *e* from *S***, since it corresponds exactly to deleting an edge from a graph. The dual operation, called the **contraction of *e* from *S***, gives a matroid M''_e on $S \backslash e$ whose independent sets are those subsets X of $S \backslash e$ such that $X \cup e$ is independent in M (except in the trivial case when e is a loop of M, in which case M''_e is the same as M'_e). Contraction is the matroid operation which corresponds to the contraction of an edge in a graph, and to the projection from a point in a vector space. The key facts about these operations are as follows:

Theorem 3.1. *The contraction and deletion of elements are commutative matroid operations—that is, for any elements e, f of M,* $(M'_e)''_f = (M''_f)'_e$. ||

Theorem 3.2. *Contraction and deletion are dual operations in the sense that* $(M'_e)^* = (M^*)''_e$, $(M''_e)^* = (M^*)'_e$. ||

We define a **minor** of M to be any matroid obtainable from M by a series of contractions and deletions, and write $M > N$ to denote that N is a minor of M.

Example 3.1. The uniform matroid $U_{k,n}$ has as its bases all k-sets of an n-set S, and its dual has as bases all $(n-k)$-subsets of S. So, $(U_{k,n})^* = U_{n-k,n}$.

Example 3.2. Let G be the plane graph of Fig. 2(a). Then $(M(G))^*$ has as its bases the complements of the spanning trees of G—namely, the sets $\{a, c\}$, $\{a, d\}$, $\{a, e\}$, $\{b, c\}$, $\{b, d\}$, $\{b, e\}$, $\{c, d\}$, $\{d, e\}$. But these sets are just the spanning trees of the graph G' of Fig. 2(b), which is in turn recognizable as the planar dual of G. As we shall see, this is just a particular case of a general result.

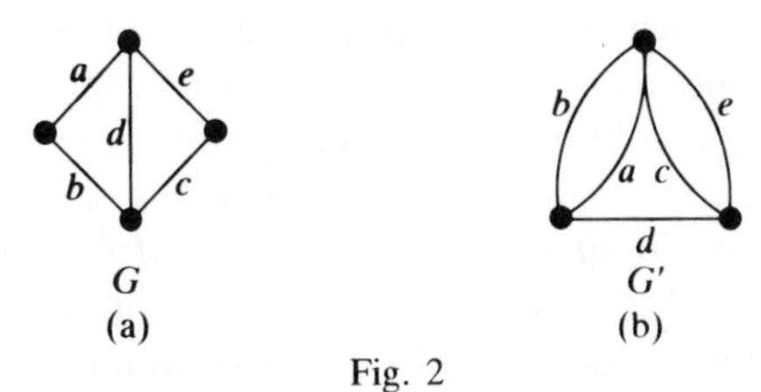

Fig. 2

To this end, let $\mathscr{C}^*$ denote the collection of cocycles of a graph G. It is not difficult to verify that $\mathscr{C}^*$ satisfies the circuit axioms of a matroid. We call this the **cocycle matroid** of G and denote it by $M^*(G)$. Furthermore, it can be shown that the bases of $M^*(G)$ are exactly those subsets Y of $E(G)$ such that $E(G) \backslash Y$ is the edge-set of a maximal spanning forest of

G. This proves the following fundamental result of Whitney:

Theorem 3.3. *For any graph G, the cycle matroid $M(G)$ and cocycle matroid $M^*(G)$ are dual. If G is planar and H is a planar graph dual to G, then $M(G) = M^*(H)$ and $M(H) = M^*(G)$.* ||

There are many matroid theorems of a fairly routine nature which interrelate these concepts. For example:

Theorem 3.4. *A subset of S is a base of a matroid M if and only if it meets every cocircuit of M and is minimal with respect to this property.* ||

This result is the matroid extension of the graph theorem that T is a maximal spanning forest of a graph G if and only if it is a minimal subset which intersects every cocycle of G.

We can also prove that, *for any circuit C and cocircuit C^* of a matroid M, $|C \cap C^*| \neq 1$.* The corresponding property in a graph is that the number of edges in the intersection of a cycle and a cocycle is even.

Note also that, in the same way that a matroid can be defined by its bases or circuits, it is also determined by its cobases or its cocircuits. This is because the cobases of M are the bases of M^*, and hence determine M^* uniquely; by the uniqueness of the dual, they must also determine M.

We emphasize that one of the most important aspects of matroid theory is that it enables us to deal with the dual of a graph even when the graph is not planar. As we shall see in Sections 7 and 9, this has attractive and important consequences. We conclude this section by introducing some terminology which is useful when discussing such questions.

A matroid is **graphic** if it is isomorphic to the cycle matroid of some graph, and **cographic** if it is isomorphic to the cocycle matroid of some graph. By Whitney's theorem, a matroid is both graphic and cographic if and only if it is isomorphic to the cycle matroid of a planar graph.

The smallest matroid which is neither graphic nor cographic is the uniform matroid $U_{2,4}$ which (for obvious reasons) is sometimes known as the 4-point line. One reason why $U_{2,4}$ is not graphic is that it fails to satisfy the following condition which clearly holds for graphic matroids: *a necessary condition for a matroid to be graphic (or cographic) is that, for each circuit C and cocircuit C^*, the cardinality of $C \cap C^*$ is even.*

There are many other necessary conditions of the same type for a matroid to be graphic or cographic. We shall summarize these in Section 5.

4. Representable Matroids

Let V be a vector space over the field $GF(q)$, and let S be any subset of V. The collection $\mathcal{I}$ of all linearly independent subsets of S clearly satisfies the independence axioms for a matroid, and any matroid which is isomorphic to such a matroid is said to be **representable** (or sometimes **coordinatizable**) over $GF(q)$. As an example, in Fig. 3 we show that the cycle matroid $M(K_4)$ is representable over any field, by showing its equivalence to a configuration of points and lines in 2-space and the equivalent columns of a matrix.

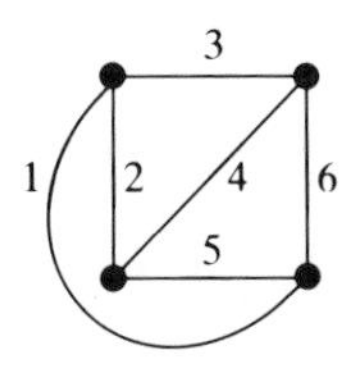

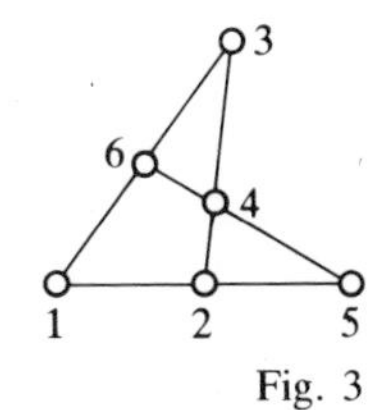

1	2	3	4	5	6
1	0	0	0	1	1
0	1	0	1	1	0
0	0	1	1	0	1

Fig. 3

Not all matroids are representable, however. For example, the following 9-element matroid is not representable over any field. Let $S = \{1, 2, \ldots, 9\}$ and let $\mathcal{I}$ consist of all subsets of cardinality at most 3, except for those which are collinear in Fig. 4. This example is not the smallest non-representable matroid; there exists an 8-element matroid of rank 4 which is not representable over any field, but the reasons for this are different.

Those familiar with projective geometry will recognize Fig. 4 as the Pappus configuration with one line missing. Since Pappus' theorem must hold in any geometrical configuration coordinatized over a field, this matroid is not representable over any field. Inserting the 'missing line' yields a matroid which is representable.

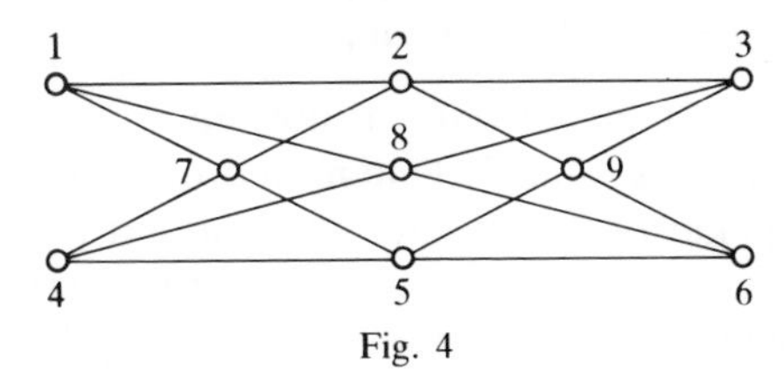

Fig. 4

A related example is shown in Fig. 5, which gives an isomorphism between the cycle matroid $M(K_5)$ and the 3-dimensional Desargues configuration.

A basic property of representable matroids is the following:

Theorem 4.1. *If a matroid M is representable over a field F, then so are its dual and all of its minors.* ||

A matroid is called **regular** if it is representable over every field. Graphic and cographic matroids have this property:

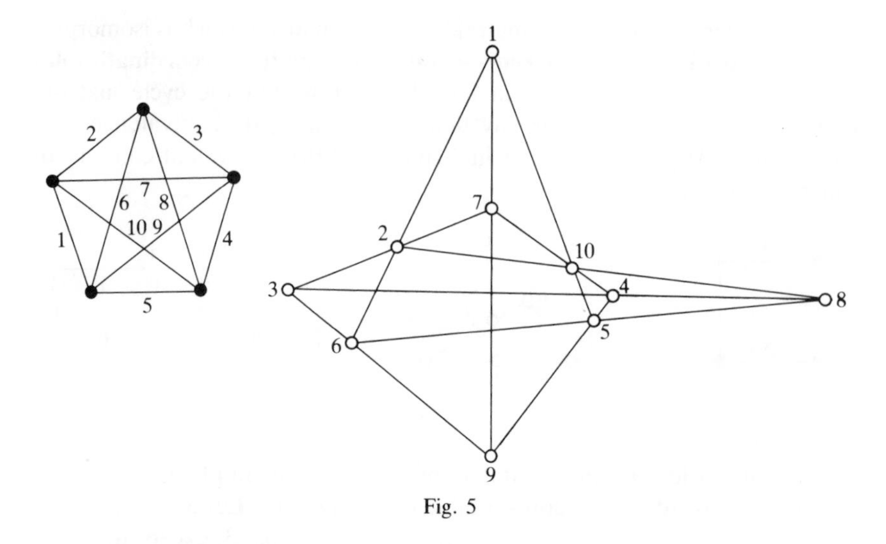

Fig. 5

Theorem 4.2. *Every graphic or cographic matroid is regular.* ||

It is not difficult to check that the uniform matroid $U_{2,4}$ is representable over every field except $GF(2)$. We call a matroid which is representable over $GF(2)$ a **binary matroid**. This definition implies that all regular matroids, and hence all graphic and cographic matroids, are binary. It is not difficult to characterize binary matroids:

Theorem 4.3. *A matroid is binary if and only if the intersection of each circuit and cocircuit has even cardinality.* ||

The smallest binary matroid which is not representable over every field is the **Fano matroid**. This is the 7-element matroid of rank 3 corresponding to the projective plane $PG(2, 2)$, and we denote it by F_7. It has a binary representation given by the columns of the first matrix below. Its dual matroid F_7^*, sometimes called the *heptahedron*, has rank 4 and has a binary representation given by the columns of the second matrix below. It is not difficult to verify that F_7 and F_7^* are representable only over fields of characteristic 2.

$$F_7 = \begin{pmatrix} 1\,0\,0\,1\,1\,0\,1 \\ 0\,1\,0\,1\,0\,1\,1 \\ 0\,0\,1\,0\,1\,1\,1 \end{pmatrix}, \quad F_7^* = \begin{pmatrix} 1\,0\,0\,0\,1\,1\,0 \\ 0\,1\,0\,0\,1\,0\,1 \\ 0\,0\,1\,0\,0\,1\,1 \\ 0\,0\,0\,1\,1\,1\,1 \end{pmatrix}$$

5. Excluded Minor Theorems

Kuratowski's characterization of planar graphs is a classic theorem of graph theory. It is the prototype of what we call *theorems of excluded-minor type*. These state that a matroid possesses a certain property if and only if it has no minor of a given finite class. Recall that we write $M > N$ to denote that N is a minor of M. A version of Kuratowski's theorem can then be stated in matroid language as follows:

Theorem 5.1. *A graphic matroid M is cographic if and only if $M \not> M(K_5)$ or $M(K_{3,3})$.* ||

Most excluded minor theorems in matroid theory have difficult proofs. The easiest is the following characterization of binary matroids by Tutte [**37**]. Recall that the 4-point line $U_{2,4}$ is the smallest matroid which is not representable over $GF(2)$. Tutte proved the following result:

Theorem 5.2. *A matroid M is binary if and only if $M \not> U_{2,4}$.* ||

Much harder is Tutte's main theorem [**37**], although a simpler (but still difficult) proof has been found by Seymour [**32**]:

Theorem 5.3 (Tutte's Theorem). *A matroid M is regular if and only if it is binary and $M \not> F_7$ or F_7^*.* ||

It is obvious that in order to be regular, a matroid must be binary and cannot contain F_7 or F_7^* as a minor since they are representable only over fields of characteristic 2. The difficult part is proving sufficiency. By Theorems 5.2 and 5.3, necessary conditions for M to be graphic are that $M \not> U_{2,4}$, F_7 or F_7^*. This is still not sufficient since, for example, $M^*(K_5)$ satisfies this condition but is not graphic (for if it were, the graph K_5 would be planar).

Tutte's second major theorem [**38**] is the following; the second result follows immediately by dualizing the first:

Theorem 5.4. (*i*) *A matroid M is graphic if and only if it is regular and $M \not> M^*(K_5)$ or M^* $(K_{3,3})$;*

(*ii*) *M is cographic if and only if it is regular and $M \not> M(K_5)$ or $M(K_{3,3})$.* ||

Alternatively, we can write for part (*i*):

Theorem 5.4. (*i*) *M is graphic if and only if $M \not> U_{2,4}$, F_7, F_7^*,* $M^*(K_5)$ or $M^*(K_{3,3})$. ||

Until the mid-1970s these results were the classics of matroid theory, combining a rare economy of statement with depth and difficulty of proof. Although sometimes described as the matroid counterpart of Kuratowski's graph theorem, these theorems are much harder to prove. However they have recently been joined at the pinnacle by another theorem.

To introduce this, we consider the problem of representing a matroid over a field other than $GF(2)$—say, $GF(3)$. First, we note that F_7 (and hence F_7^*) is representable only over fields of characteristic 2. Secondly, we note that the 5-point line $U_{2,5}$ is not coordinatizable over $GF(3)$; there is nothing subtle about this—it is just the observation that the field is too small. By duality, it follows that $U_{3,5}$ is not representable over $GF(3)$. The following remarkable result of Bixby **[5]**, R. Reid (unpublished), and Seymour **[32]** states that $U_{2,5}$ and F_7 and their duals are the only excluded minors:

Theorem 5.5. *A matroid M is representable over $GF(3)$ if and only if $M \not> U_{2,5}, U_{3,5}, F_7$ or F_7^*.* ||

No excluded minor condition is known for representability over any other fields, and it may not be unreasonable to conjecture that if $q \neq 2$ or 3, then there is no finite list of excluded minors which guarantee representability over $GF(q)$.

Since 1977, Seymour **[33]**, **[35]** has found excluded minor conditions for several other matroid properties having nothing (on the surface at least) to do with representability. It appears from these results that excluded minor theorems are far more common in matroid theory than in graph theory.

6. Transversal Theory

We mentioned in the Introduction that one of the most important early applications of matroid theory was in the field of transversals or matchings. In this section we give a brief sketch of some of the main ideas; lucid and comprehensive treatments can be found in the books by Mirsky **[24]** and Bryant and Perfect **[6]**.

We shall be restricted to finite sets, although some of the more important ideas also carry over to infinite sets (see **[24]**). If $\mathcal{A} = \{A_1, \ldots, A_m\}$ is a collection of subsets of the set S, we say that $X = \{x_1, \ldots, x_k\}$ is a **partial transversal** of $\mathcal{A}$ if there exists a one-to-one mapping $\psi: \{1, \ldots, k\} \to \{1, \ldots, m\}$ such that $x_i \in A_{\psi(i)}$. The **length** of this partial transversal is k, and if $k = m$ then we call the partial transversal a **transversal**.

The well-known theorem of P. Hall **[15]** gives necessary and sufficient conditions for a family of sets to have a transversal, and many theorems in the literature give necessary and sufficient conditions for a family of sets to have a transversal with prescribed properties. This whole area can also be regarded as a theory of matchings in bipartite graphs. From either viewpoint, the following result, which is surprisingly straightforward to prove, is of fundamental importance.

Theorem 6.1. *For any finite family $\mathscr{A}$ of subsets of a finite set S, the collection of partial transversals of $\mathscr{A}$ forms the independent sets of a matroid on S.* ||

We denote this matroid by $M(\mathscr{A})$, and call a matroid M a **transversal matroid** if it is isomorphic to $M(\mathscr{A})$ for some family $\mathscr{A}$. Relatively few matroids are transversal; the smallest matroids which are not transversal are the cycle matroids of the graphs C_3^2 and K_4 in Fig. 6.

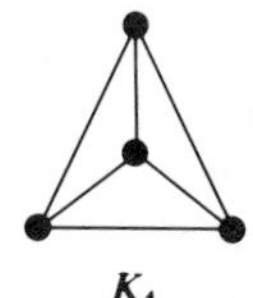

Fig. 6

Unlike the situation with graphic, binary and regular matroids, there seems to be no excluded minor condition for a matroid to be transversal. One of the reasons for this may be that the class of transversal matroids is not closed under the taking of minors; in particular, the contraction of a transversal matroid need not be transversal. Another reason is that the class of transversal matroids is not closed under the taking of duals. For example, consider the matroid $(M(C_3^2))^*$. Since C_3^2 is a planar graph, we have $(M(C_3^2))^* = M(K_{2,3})$. It is straightforward to verify that $M(K_{2,3})$ is a transversal matroid, but its dual is obviously not.

Despite the fact that the class of transversal matroids is not closed under duality, the class of 'dual transversal matroids' has an interesting graphical interpretation, which we now consider.

Recall the vertex version of Menger's theorem, which is concerned with the maximum number of vertex-disjoint paths linking two sets of vertices in a graph. Associated with such linkings in graphs is an attractive class of matroids defined as follows.

Let D be a digraph, and let A be a subset of $V(D)$. Define $\mathscr{I}_A$ to be the collection of subsets X of V which have the property that, if $X = \{x_1, \ldots, x_t\}$, there are t vertex-disjoint paths $P_1, \ldots, P_t$ having x_i as initial vertex and having the terminal vertex in A.

Theorem 6.2. *For any digraph D and any subset $A \subseteq V(D)$, the set $\mathcal{I}_A$ is the set of independent sets of a matroid $M(D; A)$ on the vertex-set $V(D)$.* ||

Any matroid M which is isomorphic to $M(D; A)$, for some D and A, is called a **strict gammoid**.

Strict gammoids bear almost the same relationship to Menger's theorem as transversal matroids do to Hall's theorem. This relationship is made precise in the following theorem:

Theorem 6.3. *A matroid is the dual of a transversal matroid if and only if it is a strict gammoid.* ||

At first sight this theorem might seem more of interest to a matroid-theorist than a graph-theorist, but it does in fact lead to considerable insight into what is going on and what makes linking theorems work. Moreover, it suggests that to almost every variant of Hall's theorem (as given, say, in Mirsky [**24**]), there is a dual version concerning the existence of linked sets of vertices. We shall not pursue this topic here but will outline some of the more important results and applications of matroids which have ramifications in transversal theory.

We first recall Hall's theorem which states:

Theorem 6.4. *If $\mathcal{A} = \{A_1, \ldots, A_m\}$ is any family of subsets of the finite set S, then $\mathcal{A}$ has a transversal if and only if, for each $J \subseteq \{1, \ldots, m\}$,*

$$\left|\bigcup_{i \in J} A_i\right| \geq |J|. \quad \|$$

One of the most fundamental theorems in transversal theory is the following matroid generalization of Hall's theorem, which was proved by Rado in 1942 and passed almost unnoticed for two decades:

Theorem 6.5 (Rado's theorem). *If M is a matroid on S, and if $\mathcal{A} = \{A_1, \ldots, A_m\}$ is any family of subsets of a set S, then $\mathcal{A}$ has a transversal which is independent in M if and only if, for each $J \subseteq \{1, \ldots, m\}$,*

$$r\left(\bigcup_{i \in J} A_i\right) \geq |J|. \quad \|$$

The importance of this theorem is that, merely by constructing 'useful matroids' on S, we can obtain many of the Hall-type theorems proved by Hoffman and Kuhn [**16**], [**17**] and others by *ad hoc* or linear programming methods. For example, if we let M be the trivial matroid in which every subset is independent, then we get Hall's theorem. By taking M to have as bases only those subsets of S having cardinality m and containing a prescribed subset E, we get the conditions for $\mathcal{A}$ to have a transversal containing the prescribed subset E.

Many other classes of matroids of the transversal-gammoid type exist. For example, *matching matroids* have as their ground set the vertex-set of a graph G, and a set X is independent if it is part of a matching in G. Details of these are given in [**47**, Chapter 14]. However, up to now the most interesting applications have been to the classes of matroids described above.

7. Unions of Matroids

We now turn to a theorem of Nash-Williams [**26**] on matroids, which is particularly interesting to graph-theorists as an example of the use of mathematical generalization. In the 1960s, Tutte [**39**] and Nash-Williams [**25**] answered the following two questions by ingenious but intricate graph-theoretic arguments:

Question 1. When does a graph G have k edge-disjoint spanning trees?

Question 2. What is the minimum number of disjoint subforests whose union is G?

These questions are just special cases of the more general questions:

Question 1′. When does a matroid M have k disjoint bases?

Question 2′. When is a set S the union of k independent sets of a matroid M on S?

We now show that these questions can be answered quite easily, using the following matroid theorem of Nash-Williams [**26**]:

Theorem 7.1. *Let $M_1, \ldots, M_k$ be matroids on the same set S. Let their respective rank functions be r_i, and let $\mathcal{I}_i$ be the family of independent sets of M_i. Let $\mathcal{I}$ be the family of subsets of S of the form $X_1 \cup \ldots \cup X_k$, where $X_i \in \mathcal{I}_i$. Then $\mathcal{I}$ is the set of independent sets of a matroid on S, denoted by $M_1 \vee M_2 \vee \ldots \vee M_k$, which has rank r given by*

$$r = \min_{A \subseteq S} [r_1(A) + \ldots + r_k(A) + |S - A|]. \; \|$$

We now show how to use this result to answer Questions 1′ and 2′. Given a matroid M, take $M_i = M$ for all $i = 1, \ldots, k$. Then, by Theorem 7.1, the matroid M has k disjoint bases if and only if the matroid $M^{(k)} = M \vee \ldots \vee M$ (k times) has rank $kr(M)$:

Corollary 7.2. *The matroid M has k disjoint bases if and only if, for each subset A of S,*

$$|S - A| \geq k(r(M) - r(A)). \; \|$$

Similarly, S is the union of k independent sets if and only if $M^{(k)}$ has rank equal to $|S|$. Applying Theorem 7.1, we get:

Corollary 7.3. *If M is a matroid on S, then S is the union of k independent sets if and only if, for all $A \subseteq S$,*

$$kr(A) \geq |A|. \; \|$$

Applying these results to a graph G, we get conditions which answer Questions 1 and 2.

What is probably most interesting about this approach to the above questions is that the proof of the more general theorem about matroids is much easier than the original proofs of the less general graph theorems of Tutte and Nash-Williams. In other words we have a classic example of 'simplification by generalization'.

8. Colorings and Flows

The chromatic polynomial of a graph is a very familiar concept in graph theory (see Chapter 2). Indeed, it has become so familiar that one tends to overlook the significance of the fact that, if we calculate the chromatic polynomial of a given graph by the well-known technique of successively contracting and deleting edges, then *the resulting polynomial is independent of the order in which we consider the edges.*

This basic observation is the key to the Tutte–Grothendieck theory developed by T.H. Brylawski [**7**] in 1972 and having wide applications in fields as far apart as percolation theory (Oxley and Welsh [**28**]) and codes (Greene [**14**]).

Brylawski's idea was to define a 'Tutte–Grothendieck' invariant for matroids as follows. A **matroid invariant** is a function f from the set of matroids to a commutative ring such that, if M is isomorphic to N, then $f(M) = f(N)$. It is a **Tutte–Grothendieck invariant** if, in addition, it satisfies the following two conditions:

(*i*) $f(M) = f(M'_e) + f(M''_e)$, for any edge e which is not a loop or coloop;

(*ii*) $f(M_1 + M_2) = f(M_1)f(M_2)$.

It is easy to check that examples of Tutte–Grothendieck invariants of matroids are the number of bases, the number of independent sets and the number of spanning sets.

Brylawski's main result is the following:

Theorem 8.1. *Any Tutte–Grothendieck invariant is uniquely determined by its values on the two single-element matroids consisting of a loop or coloop. More precisely, there exists a polynomial $T(M;\, x,\, y)$ in two* *variables x, y such that, if f is a Tutte-Grothendieck invariant, and if $f(M_0) = x$ and $f(M_0^*) = y$, where M_0 is the single-element matroid of rank 1, then*

$$f(M) = T(M; x, y). \parallel$$

The polynomial $T(M; x, y)$ is known as the **Tutte polynomial** of the matroid M. It is most easily defined by its relation to the (Whitney) rank-generating function $R(M; x, y)$:

$$R(M; x, y) = T(M; x - 1, y - 1),$$

where

$$R(M; x, y) = \sum_{A \subseteq S} x^{r(S)-r(A)} y^{|A|-r(A)}.$$

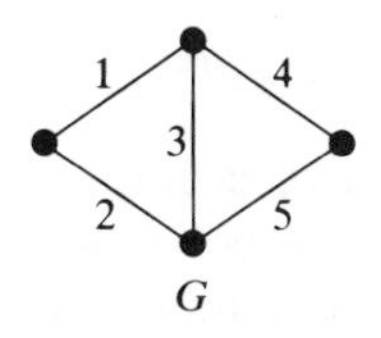

Fig. 7

For example, consider the graph G in Fig. 7, and suppose we delete and contract the edges in their numerical order. Then we get the Tutte polynomial $T(M(G); x, y)$ as follows:

$$T(M(G);x,y) = \quad + \quad$$

$$= x \quad + (\quad + \quad)$$

$$= (x+1) \quad + y \quad$$

$$= (x+1)(\quad + \quad) + y(\quad + \quad)$$

$$= (x+1)(x^2 + \quad + \quad) + y(x+y)$$

$$= x^3 + x^2 + xy + x^2 + x + y + xy + y^2$$

$$= x^3 + 2x^2 + 2xy + y^2 + x + y,$$

where, in the obvious notation, we use x for coloops (=bridges) and y for loops.

The remarkable thing is that, no matter in what order we carry out our contractions and deletions, we always get the same polynomial. Furthermore, any function which can be calculated by this contraction–deletion method must be an evaluation of this polynomial for suitable x and y. For example, $b(M)$, the number of bases of M, is given by $b(M) = T(M; 1, 1)$, since the value of b on a coloop and a loop is in both cases 1. Similarly, $i(M)$, the number of independent sets of M, is given by $i(M) = T(M; 2, 1)$, since a coloop has two independent sets whereas a loop has only one.

The prototype Tutte–Grothendieck invariant is the chromatic polynomial of a graph. It can be defined for a matroid as the **chromatic polynomial**

$$P(M; \lambda) = (-1)^{r(S)} T(M; 1 - \lambda, 0).$$

When M is the cycle matroid of a graph G, we know from elementary graph theory that the chromatic polynomial $P(G; \lambda)$ of the graph G is essentially another Tutte–Grothendieck invariant, since it satisfies the relation

$$P(G; \lambda) = P(G'_e; \lambda) - P(G''_e; \lambda),$$

when e is neither a bridge nor a loop of G. The analogue of condition (*ii*) at the beginning of the section is trivial, and it is straightforward to see that

$$\lambda^{k(G)} P(M(G); \lambda) = P(G; \lambda),$$

where $k(G)$ is the number of connected components of G. Accordingly, if we define the **chromatic number** $\chi(M)$ of a matroid by

$$\chi(M) = \inf_{k \in Z^+} \{k: P(M; k) > 0\},$$

then the chromatic number of a graphic matroid is the same as the chromatic number of the underlying graph. Incidentally, this shows that two graphs whose cycle matroids are isomorphic must have the same chromatic number. For a derivation and proof of all these results see Welsh [**47**, Chapter 15].

The Flow Polynomial

An easily verifiable property of Tutte polynomials is that

$$T(M; x, y) = T(M^*; y, x),$$

and so it is natural to ask whether the chromatic polynomial of the cocycle matroid of a graph has any natural interpretation. We now turn our attention to this problem.

Suppose that G is a graph, and that ω is any orientation of the edges of G. If H is any Abelian group, we say that a mapping φ: $E(G) \to H\backslash\{0\}$ (where 0 is the zero of H) is an ***H*-flow** if, for each vertex v of G,

$$\sum_{\partial^+(v)} \varphi(e) - \sum_{\partial^-(v)} \varphi(e) = 0,$$

where $\partial^+(v)$ and $\partial^-(v)$ denote, respectively, the edges of G which are oriented into and out of v in the orientation ω. (Note that what we call an H-flow is called a *nowhere zero H-flow* by most authors.) If ω and ω' are any two orientations of G, then there is an obvious one-to-one correspondence between flows in ω and ω'. Thus, the number $N(H; G)$ of H-flows in G is independent of its orientation, and is a function only of H and G. Thus, we speak henceforth of an undirected graph G as **supporting** an H-flow if an H-flow is possible for any orientation of G.

Secondly (and this is a very nice application of the Tutte–Grothendieck theory), we have the following theorem; its proof is a straightforward application of the contraction–deletion technique:

Theorem 8.2. *The number of H-flows on a graph G depends only on the order of the group H, and is given by evaluating the chromatic polynomial of the cocycle matroid $M^*(G)$ at $\lambda = o(H)$, the order of the group H.* ||

As an example, we consider H-flows on any directed version of K_4. Since $M^*(K_4)$ has chromatic polynomial $(\lambda - 1)(\lambda - 2)(\lambda - 3)$, the number of $(Z_2 \times Z_2)$-flows on K_4 is $3 \times 2 \times 1 = 6$.

Thus we may sensibly say that **a graph *G* has a *k*-flow** if, for any orientation of G and any Abelian group H of order k, there exists an H-flow on G. Moreover (and this is most crucial), we can decide whether or not a particular graph supports a k-flow by calculating the chromatic polynomial of its cocycle matroid and then evaluating this polynomial at $\lambda = k$. As a first consequence of Theorem 8.2, we have the following result:

Theorem 8.3. *Any bridgeless planar graph G has a 4-flow.*

Proof. This statement follows from the four-color theorem of Appel and Haken [2]. To see this, we note that, if G is planar and bridgeless, then there exists a (dual) graph G^* which is planar and loop-free such that $M^*(G) = M(G^*)$.

Hence

$$P(M^*(G); 4) = P(M(G^*); 4),$$

and by the four-color theorem we know that

$$P(M(G^*); 4) > 0,$$

since G^* is planar. ||

In his fundamental paper in 1954, Tutte **[36]** made two conjectures. The first, that there exists some integer n such that any bridgeless graph G has an n-flow, was settled by Jaeger **[19]** who showed that any bridgeless graph G has an 8-flow. This nice result has been improved by Seymour **[34]**, who has proved the following theorem:

Theorem 8.4. *Every bridgeless graph has a 6-flow.* ||

The second conjecture of Tutte **[36]** is still unsettled, and is known as the *5-flow conjecture*; it can be stated as follows:

Conjecture 8.1 (Tutte's 5-flow conjecture). *Every bridgeless graph has a 5-flow.*

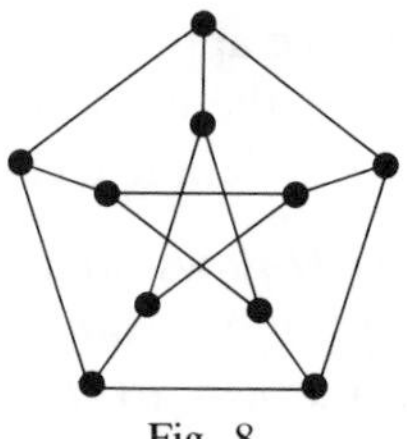

Fig. 8

To see that this conjecture, if true, is best possible consider the Petersen graph P shown in Fig. 8. It is non-trivial to verify that P has no k-flow for $k \leq 4$, and that it is the smallest bridgeless graph with this property. Alternatively, it can be shown (with sufficient patience) that

$$P(M^*(\mathrm{P}); \lambda) = (\lambda - 1)(\lambda - 2)(\lambda - 3)(\lambda - 4)(\lambda^2 - 5\lambda + 10),$$

and thus P has no 4-flow.

The situation is somewhat different for trivalent graphs. If G is a bridgeless trivalent graph, then it is not difficult to verify that *G has a 4-flow if and only if G is 3-edge-colorable*. Using this result, we can formulate a second conjecture of Tutte **[36]**, as follows:

Conjecture 8.2 (Tutte's 4-flow conjecture). *A trivalent bridgeless graph G has a 4-flow if it has no subgraph contractible to the Petersen graph.*

9. From Graphs to Matroids

As noted earlier, much of the language and many ideas of matroid theory have been motivated by generalizing ideas in graphs to these more abstract structures. In this section, we sketch the ideas behind what is probably the most important application of matroids.

First, we make a few brief remarks on connectivity. There is nothing in matroid theory which exactly corresponds to connectivity in a graph. For instance, any two forests with the same number of edges have the same cycle matroids. However, there is a concept of k-connectivity for matroids which corresponds to a form of edge-connectivity for graphs.

We say that a matroid M on S has a ***k*-separation** if there exists $A \subseteq S$ with

$$|A| \geq k, \ |S \backslash A| \geq k \text{ and } r(A) + r(S \backslash A) \leq r(S) + k - 1.$$

It is not difficult to verify that the cycle matroid $M(G)$ of a graph G has a 1-separation if and only if G is not a 2-connected graph. Similarly, $M(G)$ has a 2-separation but not a 1-separation if and only if G has connectivity 2. (For higher values of k, graphic interpretations have been given by Tutte.) We define a matroid M to be **connected** if M has no 1-separation. Thus, matroid connectedness corresponds to 2-connectivity in graphs. Equivalently, we can see that M is disconnected if and only if it is the direct sum of smaller matroids.

Now suppose that $\mathcal{F}$ is a class of binary matroids which is closed under the taking of minors. We say that the matroid $N \in \mathcal{F}$ is a **splitter** for $\mathcal{F}$ if every M in $\mathcal{F}$ with a proper minor isomorphic to N has a 1-separation or a 2-separation. The idea of splitters comes from the matroid generalizations of the well-known graph-coloring theorem of Wagner [**45**]. The first matroid splitter in the literature (although it was not described as such) was V_8, the 8-vertex Möbius strip, shown in Fig. 9.

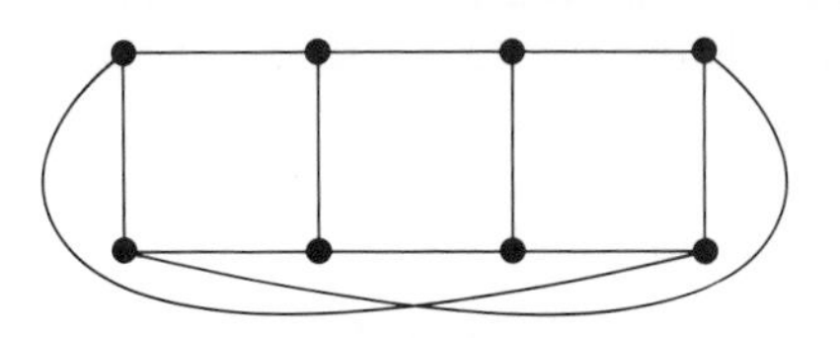

Fig. 9

Wagner's proof of the equivalence of Hadwiger's conjecture with the four-color conjecture hinged on the fact that $M(V_8)$ is a splitter for the class of graphic matroids with no minor isomorphic to $M(K_5)$.

Knowledge of a splitter for a class of matroids or graphs gives considerable leverage to inductive proofs, since we either know the matroid exactly or can break it up into smaller submatroids. The trouble is that splitters for a class of matroids are hard to find, even when they exist. There seems to be no routine way to test whether a given class has a splitter or to find a splitter if one exists. It is, however, straightforward to verify that a given matroid N is a splitter, as we now show.

Let $\mathscr{F}$ be a class of binary matroids which is closed under the taking of minors, and denote by $\mathscr{F}^*$ the class of duals of the matroids in $\mathscr{F}$. We say that a matroid $N \in \mathscr{F}$ is **compressed in** $\mathscr{F}$ if

(*i*) N is simple (that is, N has no circuit of size at most 2), and

(*ii*) if $M \in \mathscr{F}$ and $M'_e = N$, then e is a loop or coloop or is parallel to some other element of M.

Using this idea, one can prove the following result:

Theorem 9.1. *Let $\mathscr{F}$ be a class of binary matroids which is closed under minors and isomorphism, and suppose that $N \in \mathscr{F}$. If N is non-null, connected, and satisfies the conditions*

(i) N is compressed in $\mathscr{F}$;

(ii) N^* is compressed in $\mathscr{F}^*$;

(iii) N is not isomorphic to the cycle matroid $M(W_p)$ of the wheel graph W_p, for any $p \geq 3$;

then N is a splitter for $\mathscr{F}$. ||

Using this result, one can, with sufficient time, check that F_7^* is a splitter for the class of binary matroids with no F_7-minor. Similarly, one can use Theorem 9.1 to show that the matroid R_{10}, consisting of the ten 5-vectors over $GF(2)$ with exactly three ones, is a splitter for the class of regular matroids.

To appreciate fully Seymour's main theorem, we need one last set of definitions. A **cycle** of a matroid is a union of disjoint circuits. If M_1 and M_2 are binary matroids on S_1 and S_2, respectively, we define $M_1 \Delta M_2$ to be the matroid on the symmetric difference $S_1 \Delta S_2$ whose cycles are those subsets of $S_1 \Delta S_2$ of the form $C_1 \Delta C_2$, where C_i is a cycle of M_i. Now assume that $|S_1| < |S_1 \Delta S_2|$ and $|S_2| < |S_1 \Delta S_2|$:

(*i*) if $S_1 \cap S_2 = \emptyset$, we call $M_1 \Delta M_2$ a **1-sum** of M_1 and M_2;

(*ii*) if $S_1 \cap S_2 = \{z\}$, where z is not a loop or coloop of M_1 or M_2, we call $M_1 \Delta M_2$ a **2-sum** of M_1 and M_2;

(*iii*) if $|S_1 \cap S_2| = 3$ and $S_1 \cap S_2 = Z$, where Z is a circuit of M_1 and M_2, we call $M_1 \Delta M_2$ a **3-sum** of M_1 and M_2.

In each case we call M_1 and M_2 the **parts** of the sum.

Now a 1-sum is just the direct sum of graph theory, and the 2-sum and 3-sum are the matroid operations corresponding to 'sticking' two graphs together along an edge or triangle, respectively, and then deleting the edge or triangle in question. In either case, if we have a matroid M which is the 2-sum or 3-sum of other matroids, then we have a convenient way of breaking M up into these smaller structures.

Seymour's main theorem is the following remarkable characterization of regular matroids:

Theorem 9.2. *If M is a regular matroid, then M is the* 1-*sum*, 2-*sum or* 3-*sum of graphic matroids, cographic matroids and copies of the* 10-*element matroid* R_{10}. ||

The power of this result has not yet been fully exploited. However, as one major example, we show in the next section how it gives 'as a corollary' a major breakthrough on a long-standing algebraic coloring problem of Tutte [**36**] and relates it to the flow problem discussed in the last section.

More importantly, since we know that a binary matroid is regular if and only if it has a matrix representation which is totally unimodular, Seymour's result on regular matroids has proved to be a major step in understanding what gives these matrices their attractive properties in combinatorics and optimization.

10. The Blocking Problem

In this section we use matroids to relate the fundamental graph coloring and flow problems of Section 8 to a blocking problem in projective spaces which goes back at least as far as Veblen [**44**] in 1912.

Consider the projective space $PG(r-1, q)$. For any positive integer t, a **t-block** in $PG(r-1, q)$ is a set X of points such that $X \cap F \neq \emptyset$ for each flat F of rank $r-t$. In particular, the 1-blocks are the sets which have non-empty intersection with every hyperplane of $PG(r-1, q)$. It follows that if X is a t-block, and if $Y \supseteq X$, then Y is a t-block. We call X a **minimal t-block** if X is a t-block but $X \backslash \{p\}$ is not a t-block for each $p \in X$.

Standard vector space arguments can be used to show that the projective space $PG(t, q)$ is a minimal t-block over the field $GF(q)$, for any integer $t \geq 2$ and any prime power q. For example, $PG(2, q)$ $(=F_7)$ is a 2-block over $GF(2)$, and is a 1-block over $GF(4)$. More generally we have the following result:

Theorem 10.1. *A t-block over $GF(p)$ is a 1-block over $GF(p^t)$ for each positive integer t.* ||

The converse of Theorem 10.1 is not always true since there exist 1-blocks over $GF(p^t)$ which, when regarded as geometrical configurations, are not representable (that is, coordinatizable) over $GF(p)$. However, we can show that if a 1-block over $GF(p^t)$ is representable over $GF(p)$, then it is a t-block over $GF(p)$. This follows almost immediately from the following remarkable theorem which gives the relationship between blocking sets and the chromatic polynomial of the underlying matroid:

Theorem 10.2. *Let M be a matroid of rank r on S which is embeddable in $V(r, q)$. Then there exists an $(r - t)$-dimensional subspace F of $V(r, q)$ with $F \cap S = \emptyset$ if and only if $P(M; q^t) > 0$.* ||

The first point to notice about this theorem is that its conclusion does not depend on the embedding, but says that such a subspace exists for each embedding. In fact, the full version of Theorem 10.2 proved by Crapo and Rota **[9]** shows that $P(M; q^t)$ enumerates the collections of hyperplanes whose intersection is a flat of the type required.

The relationship with blocking is now obvious:

Theorem 10.3. *A matroid M which is representable over $GF(q)$ is a t-block over $GF(q)$ if and only if $P(M; q^t) = 0$.*

Conversely, the t-blocks over $GF(q)$ are exactly those matroids representable over $GF(q)$ for which $P(M; q^t) = 0$. ||

This statement illustrates precisely the slight abuse of language in the statement "M is a t-block". What we mean is that any of the various vector representations of M in $V(r, q)$ is a t-block; in other words, it is not their coordinatization in $V(r, q)$ which is important, but their geometrical structure.

As an example, consider the Fano matroid F_7 consisting of the seven non-zero vectors of $V(3, 2)$. Since

$$P(F_7; \lambda) = (\lambda - 1)(\lambda - 2)(\lambda - 4),$$

F_7 is a 2-block over $GF(2)$ and a 1-block over $GF(4)$. Similarly, since the uniform matroid $U_{2,n}$ has chromatic polynomial $\lambda^2 - n\lambda + n - 1$, we know that the $(q + 1)$-point line is a 1-block over the field $GF(q)$ for each prime power q.

For complete graphs K_n, we have

$$P(M(K_n); \lambda) = (\lambda - 1)(\lambda - 2) \ldots (\lambda - n + 1).$$

It follows that for any prime power $q = p^t$, the cycle matroid of K_{q+1} is a 1-block over $GF(q)$, and is thus also a t-block over $GF(p)$. Now the cycle matroid $M(G)$ of any graph G is representable over every field, and so *$M(G)$ is a t-block over $GF(q)$ if and only if G has chromatic number $\chi(G) > q^t$.* Thus the graphic 1-blocks over $GF(2)$ are just the cycle matroids of non-bipartite graphs.

When we come to 2-blocks over $GF(2)$, however, the situation is much more complicated. Since K_5 is a minimal graph which is not 4-colorable, we deduce that $M(K_5)$ *is a minimal 2-block*. However, there are many other minimal 2-blocks, since if G is any edge-critical graph for which $\chi(G) = 5$, but $\chi(G \backslash e) = 4$ for each edge e, then $M(G)$ is a minimal 2-block. An infinite family of such graphs can easily be constructed. In fact, starting with such minimal 2-blocks, both graphic and non-graphic, Oxley **[27]** has constructed other minimal 2-blocks by 'sticking them together' in a non-trivial fashion using a method called *series connection*. (This is the matroid analogue of forming the 'Hajós-union' of two graphs.)

As mentioned earlier, any edge-critical 5-chromatic graph has a cycle matroid which forms a minimal 2-block. However, each of the matroids $M(K_5)$, F_7 and $M^*(\mathrm{P})$ has the additional property that no loop-free minor is also a 2-block. A 2-block with this property is called a **tangential 2-block**. (This is not Tutte's original definition, but can be seen to be equivalent to it—see Welsh **[49]**.) In 1966, Tutte proved that these three matroids are the only tangential 2-blocks with rank at most 6, and in 1976, Datta used a complicated geometrical argument to show there is no tangential 2-block of rank 7.

Tutte's tangential 2-block conjecture, originally made in 1966 and still unsettled, can be stated in the following form:

Conjecture 10.1 (Tutte's tangential block conjecture). *The only tangential 2-blocks are $M(K_5)$, F_7 and $M^*(\mathrm{P})$.*

Seymour's theory of splitters is a major step towards proving this conjecture. We first consider Hadwiger's conjecture which, in its full form, reads as follows:

Conjecture 10.2 (Hadwiger). *If a loopless graph G is not n-colorable, then it contains a subgraph contractible to K_{n+1}.*

Dirac **[13]** showed that Hadwiger's conjecture is true for $n = 3$, and Wagner **[45]** showed that for $n = 4$ it is equivalent to the four-color conjecture, and it therefore holds for $n = 4$. Thus we know that there can be no new tangential 2-block which is the cycle matroid of a graph. Seymour **[35]** used his characterization of regular matroids to prove the following striking result:

Theorem 10.4. *Any new tangential 2-block must be the cocycle matroid of a graph.*

In other words, Seymour's result shows that Tutte's tangential 2-block conjecture is exactly equivalent to the 4-flow conjecture. Thus he has reduced this seemingly intractable geometrical problem to the conceptually much simpler problem of characterizing those graphs which have no 4-flow. More precisely, there exists a tangential 2-block other than $M(K_5)$,

F_7 and $M^*(\mathrm{P})$ if and only if there is a bridgeless graph G which does not contain a subgraph contractible to the Petersen graph and which has no 4-flow.

Outline of proof of Theorem 10.4. First assume that M is a tangential 2-block which is not $M(K_5)$, F_7 or $M^*(\mathrm{P})$. If M were graphic, then there would exist a graph G which is not contractible to K_5 and yet is not 4-colorable. This would contradict the known truth of Hadwiger's conjecture for the case $n = 4$. Hence, the only tangential 2-block which is graphic is $M(K_5)$. Now consider the existence of a non-regular tangential 2-block M_0. Since M_0 is not regular, it must contain either F_7 or F_7^* as a minor. But because F_7 is a tangential 2-block, by minimality this minor must be F_7^*. Hence, since F_7^* is a splitter for binary matroids with no F_7-minor, we know that either $M_0 = F_7^*$ or M_0 has a 2-separation. But it is not difficult to show that a tangential 2-block cannot have a 2-separation, and hence $M_0 = F_7^*$. However, $\chi(F_7^*) = 2$ and therefore F_7^* is not a tangential 2-block. So we have shown that there are no non-regular tangential 2-blocks. It remains to show that there exists no tangential block which is regular but neither graphic nor cographic.

Suppose such a matroid M exists. Then, by the decomposition theorem, it must be possible to express M as a 1-sum, 2-sum or 3-sum of graphic or cographic matroids or copies of R_{10}. An induction argument shows that this is impossible unless M is cographic, and hence the only tangential blocks which are not cographic are F_7 and $M(K_5)$. ||

The conjectures and problems posed so far seem to be hard, at least in the sense that they have stood the test of time. We conclude with a conjecture which I first made in 1980 and which may be easier to settle in the negative, but which if true would imply or further relate some of the earlier conjectures.

Conjecture 10.3. *If M is binary and has no minor isomorphic to the 3-dimensional Desargues configuration, then $\chi(M) \leq 5$.*

Note that we have presented this conjecture in its geometrical form, although, as we saw in Section 4, the matroid arising from the 3-dimensional Desargues configuration is identical to $M(K_5)$.

The motivation for this conjecture is a paper of P. N. Walton and myself [**46**] in which the following results are proved:

Theorem 10.5. (*i*) *If M is binary and has no minor isomorphic to the 3-dimensional Desargues configuration or to $PG(2, 2)$, then $\chi(M) \leq 6$;*

(*ii*) *if, in addition, Tutte's 5-flow conjecture is true, then $\chi(M) \leq 5$.* ||

References

1. M. Aigner, *Combinatorial Theory*, Springer, Berlin, 1979; *MR***80h**:05002.
2. K. Appel and W. Haken, Every planar map is four-colorable, *Bull. Amer. Math. Soc.* **82** (1976), 711–712; *MR***54**#12561.
3. C. Berge, *Graphs and Hypergraphs*, North-Holland, Amsterdam, 1973; *MR***50**#9640.
4. G. Birkhoff, Abstract linear dependence in lattices, *Amer. J. Math.* **57** (1935), 800–804.
5. R. E. Bixby, Kuratowski's and Wagner's theorems for matroids, *J. Combinatorial Theory (B)* **22** (1977), 31–53; *MR***58**#21722.
6. V. Bryant and H. Perfect, *Independence Theory in Combinatorics*, Chapman and Hall, London, 1980; *MR***82d**:05001.
7. T. H. Brylawski, A decomposition for combinatorial geometries, *Trans. Amer. Math. Soc.* **171** (1972), 235–282; *MR***46**#8869.
8. T. H. Brylawski and D. G. Kelly, *Matroids and Combinatorial Geometries*, University of North Carolina Press, 1980; *MR***81f**:05053.
9. H. H. Crapo and G.-C. Rota, *On the Foundations of Combinatorial Theory: Combinatorial Geometries*, MIT Press, Cambridge, Mass., 1970; *MR***45**#74.
10. B. T. Datta, On tangential 2-blocks, *Discrete Math.* **15** (1976), 1–22; *MR***53**#5333.
11. B. T. Datta, Nonexistence of six-dimensional tangential 2-blocks, *J. Combinatorial Theory (B)* **21** (1976), 171–193; *MR***55**#7833.
12. G. A. Dirac, A property of 4-chromatic graphs and some remarks on critical graphs, *J. London Math. Soc.* **27** (1952), 85–92; *MR***13**–572f.
13. G. A. Dirac, A theorem of R. L. Brooks and a conjecture of H. Hadwiger, *Proc. London Math. Soc.* **7** (1957), 161–195; *MR***19**–161b.
14. C. Greene, Weight enumeration and the geometry of linear codes, *Studies Appl. Math.* **55** (1976), 119–128; *MR***56**#5335.
15. P. Hall, On representatives of subsets, *J. London Math. Soc.* **10** (1935), 26–30.
16. A. J. Hoffman and H. W. Kuhn, On systems of distinct representatives, *Linear Inequalities and Related Systems*, *Ann. Math. Studies* **38**, Princeton, 1956, pp. 199–206; *MR***18**–416.
17. A. J. Hoffman and H. W. Kuhn, Systems of distinct representatives and linear programming, *Amer. Math. Monthly* **63** (1956), 455–460; *MR***18**–370.
18. F. Jaeger, On nowhere-zero flows in multigraphs, *Proc. Fifth British Combinatorial Conference* (ed. C. St. J. A. Nash-Williams and J. Sheehan), *Congressus Numerantium XV*, Utilitas Math., Winnipeg, 1976, pp. 373–378; *MR***52**#16570.
19. F. Jaeger, Flows and generalized coloring theorems in graphs, *J. Combinatorial Theory (B)* **26** (1979), 205–216; *MR***81j**:05059.
20. E. L. Lawler, *Combinatorial Optimization: Networks and Matroids*, Holt, Rinehart and Winston, New York, 1976; *MR***55**#12005.
21. S. MacLane, Some interpretations of abstract linear dependence in terms of projective geometry, *Amer. J. Math.* **58** (1936), 236–240.
22. S. MacLane, A combinatorial condition for planar graphs, *Fund. Math.* **28** (1937), 22–32.
23. S. MacLane, A lattice formulation for transcendence degrees and p-bases, *Duke Math. J.* **4** (1938), 455–468.
24. L. Mirsky, *Transversal Theory*, *Math. in Science and Engineering* **75**, Academic Press, London, 1971; *MR***44**#87.
25. C. St. J. A. Nash-Williams, Edge-disjoint spanning trees of finite graphs, *J. London Math. Soc.* **36** (1961), 445–450; *MR***24**#A3087.
26. C. St. J. A. Nash-Williams, An application of matroids to graph theory, *Theory of Graphs International Symposium (Rome)*, Dunod, Paris, 1966, pp. 263–265.
27. J. G. Oxley, Colouring, packing and the critical problem, *Quart. J. Math. (Oxford) (2)* **29** (1978), 11–22; *MR***58**#21734.

28. J. G. Oxley and D. J. A. Welsh, The Tutte polynomial and percolation, *Graph Theory and Related Topics*, Academic Press, 1979, pp. 329–339; *MR*81a:05031.
29. R. Rado, A theorem on independence relations, *Quart. J. Math. (Oxford) (2)* **13** (1942), 83–89; *MR*4–269c.
30. R. Rado, Note on independence functions, *Proc. London Math. Soc. (3)* **7** (1957), 300–320; *MR*19–522b.
31. G.-C. Rota, On the foundations of combinatorial theory, I: Theory of Möbius functions, *Z. Wahrsch.* **2** (1964), 340–368; *MR*30#4688.
32. P. D. Seymour, Matroid representation over *GF*(3), *J. Combinatorial Theory (B)* **26** (1979), 159–173; *MR*80k:05031.
33. P. D. Seymour, Decomposition of regular matroids, *J. Combinatorial Theory (B)* **28** (1980), 305–359; *MR*82j:05046.
34. P. D. Seymour, Nowhere-zero 6-flows, *J. Combinatorial Theory (B)* **30** (1981), 130–135; *MR*82j:05079.
35. P. D. Seymour, On Tutte's extension of the four-colour problem, *J. Combinatorial Theory (B)* **31** (1981), 82–94; *MR*83d:05047.
36. W. T. Tutte, A contribution to the theory of chromatic polynomials, *Canad. J. Math.* **6** (1954), 80–91; *MR*15–814c.
37. W. T. Tutte, A homotopy theory for matroids I, II, *Trans. Amer. Math. Soc.* **88** (1958), 144–174; *MR*21#336.
38. W. T. Tutte, Matroids and graphs, *Trans. Amer. Math. Soc.* **90** (1959), 527–552; *MR*21#337.
39. W. T. Tutte, On the problem of decomposing a graph into *n* connected factors, *J. London Math. Soc.* **36** (1961), 221–230; *MR*25#3858.
40. W. T. Tutte, A theory of 3-connected graphs, *Nederl. Akad. Wetensch. Proc. (A)* **64** and *Indag. Math.* **23** (1961), 441–455; *MR*25#3517.
41. W. T. Tutte, On the algebraic theory of graph coloring, *J. Combinatorial Theory* **1** (1966), 15–50, *MR*33#2573.
42. W. T. Tutte, Projective geometry and the 4-color problem, *Recent Progress in Combinatorics* (ed. W. T. Tutte), Academic Press, New York, 1969, pp. 199–207; *MR*40#5500.
43. W. T. Tutte, *Introduction to the Theory of Matroids*, American Elsevier, New York, 1971; *MR*43#1865.
44. O. Veblen, An application of modular equations in analysis situs, *Ann. Math.* **14** (1912), 86–94.
45. K. Wagner, Beweis einer Abschwächung der Hadwiger-Vermutung, *Math. Ann.* **153** (1964), 139–141; *MR*28#3416.
46. P. N. Walton and D. J. A. Welsh, On the chromatic number of binary matroids, *Mathematika* **27** (1980), 1–9; *MR*81m:05060.
47. D. J. A. Welsh, *Matroid Theory*, London Math. Soc. Monographs **8**, Academic Press, London, 1976; *MR*55#148.
48. D. J. A. Welsh, Colouring problems and matroids, *Surveys in Combinatorics* (Proc. Seventh British Combinatorial Conference), Cambridge University Press, 1979, pp. 229–257; *MR*81g:05050.
49. D. J. A. Welsh, Colourings, flows and projective geometry, *Nieuw Arch. Wisk. (3)* **28** (1980), 159–176; *MR*82h:05002.
50. H. Whitney, On the abstract properties of linear dependence, *Amer. J. Math.* **57** (1935), 509–533.
51. R. J. Wilson, *Introduction to Graph Theory*, 3d edn., Longman, Harlow, 1985; *MR*50#9643; **80g**:05028.

4
Nowhere-zero Flow Problems

FRANÇOIS JAEGER

1. Introduction

The concept of a *flow in a graph* is a useful model in Operations Research, and is also essentially identical to the concept of a current in an electrical network. It is thus not surprising that the study of flows is a classical and important topic in graph theory, which leads to rich developments and generalizations in combinatorial optimization, polyhedral combinatorics and matroid theory.

At first sight it seems that the dual concept of *tension* (or potential difference) has less importance in the literature, and appears mainly in scheduling and shortest-path problems. However, as observed by Tutte, the whole theory of vertex-colorings of graphs can be formulated in terms of tension, and this is indeed an essential part of graph theory which is intimately related to the history of its development.

In the case of planar graphs, the duality between flow and tension corresponds to the geometric duality of graphs represented in the plane. This allows a reformulation of face-coloring properties of plane graphs in terms of flow properties. For instance, one can show that the four-color theorem is equivalent to the following statement:

Every bridgeless planar directed graph has an integer flow with all edge-values in the set $\{\pm 1, \pm 2, \pm 3\}$.

GRAPH THEORY, 3
ISBN 0-12-086203-4

This led Tutte to consider similar properties for arbitrary graphs. For instance, he proposed the following *5-flow conjecture*:

Every bridgeless directed graph has an integer flow with all edge-values in the set $\{\pm 1, \pm 2, \pm 3, \pm 4\}$.

The study of this kind of problem involves an extension of the usual concept of flow, where the set of flow values is an arbitrary Abelian group. Using this extended concept, it is possible to unify such problems as Tutte's 5-flow conjecture, the cycle double cover conjecture (*every bridgeless graph has a family of cycles which together cover each edge twice*) and Fulkerson's conjecture (*every bridgeless trivalent graph has six 1-factors which together cover each edge twice*) into a single framework—the class of **nowhere-zero flow problems**.

The interest of such an approach is that it brings together different methods developed independently for various conjectures, and can also help us to formulate new and pertinent problems. The present survey lays more emphasis on the unity of the different nowhere-zero flow problems than on their specific aspects.

In Section 2 we introduce the necessary definitions and notation. We then present nowhere-zero k-flows in Section 3, and discuss the main results and conjectures on the existence of such flows in Section 4. Section 5 is devoted to the cycle double cover conjecture, Section 6 deals with Fulkerson's conjecture, and in Section 7 we present a conjecture which implies the previous ones. Some results on special classes of graphs are reviewed in Section 8, and the main contributions of the reduction methods are outlined in Section 9. We conclude, in Section 10, by mentioning some relationships with other research topics.

2. Definitions and Notation

Our definition of a graph allows loops and multiple edges. For convenience, we shall not distinguish a graph G from the various digraphs which can be obtained from G by assigning an orientation to each edge of G. This motivates the following terminology: if, for each edge of a graph G, we distinguish one initial end and one terminal end, we obtain a directed graph which will be called an **orientation** of G.

If G is a directed graph, and if $S \subseteq V(G)$, we denote by $\omega^+(S)$ the set of edges with initial end in S and terminal end not in S. We write $\omega^-(S) = \omega^+(V(G) - S)$ and $\omega(S) = \omega^+(S) \cup \omega^-(S)$. A k-subset of $E(G)$ of the form $\omega(S)$, where S is a proper non-empty subset of $V(G)$, is called a $\boldsymbol{k}$**-cut** of G. Thus G is k-edge-connected if and only if it has no l-cuts for $l < k$, and a bridge is an edge which forms a 1-cut.

If A is an Abelian group (with additive notation), and if G is a directed graph, then an **A-flow** of G is a mapping φ from $E(G)$ to A such that:

$$\text{for all } S \subseteq V(G), \quad \sum_{e \in \omega^+(S)} \varphi(e) - \sum_{e \in \omega^-(S)} \varphi(e) = 0. \tag{1}$$

The usual concept of flow corresponds to the case where A is $\mathbf{Z}$ or $\mathbf{R}$.

We make the following remarks:

(*i*) It is easy to see that the mapping φ is an A-flow if and only if equation (1) is satisfied for all sets S consisting of a single vertex of G.

(*ii*) It follows from (1) that an A-flow takes the value zero on each bridge.

(*iii*) Suppose we change the orientation of the edge e in G, and simultaneously replace $\varphi(e)$ by $-\varphi(e)$. Then (1) is still valid for all subsets S of $V(G)$, and hence the new mapping φ' is an A-flow in the new directed graph.

(*iv*) If each element of A is its own opposite—for instance, if $A = \mathbf{Z}_2{}^k$ for some $k \geq 1$—the situation is simpler. Condition (1) can then be rewritten as:

$$\text{for all } S \subseteq V(G), \quad \sum_{e \in \omega(S)} \varphi(e) = 0,$$

and this is clearly independent of the orientation of G.

The **support** $\sigma(\varphi)$ of the A-flow φ of G is the set of edges e of G such that $\varphi(e) \neq 0$. φ is said to be a **nowhere-zero flow** if $\sigma(\varphi) = E(G)$. If φ takes all its values in $B \subseteq A$, then it is called a **B-flow**. We shall be interested in the existence of B-flows for subsets B for which $0 \notin B$ and $B = -B$.

Note that the problem of the existence of a B-flow in a graph is trivial if $0 \in B$ (consider the flow which takes the value 0 on every edge). On the other hand, the condition $0 \notin B$ implies that we are restricting our attention to bridgeless graphs (see remark (*ii*)). Also, by remark (*iii*), the condition $B = -B$ implies that the following properties are equivalent for a graph G:

(*a*) some orientation of G has a B-flow;

(*b*) every orientation of G has a B-flow.

When (*a*) and (*b*) hold, we simply say that *G has a B-flow*. Thus we are studying a property of undirected graphs. The orientations will be used only as a reference for defining flows. By remark (*iv*), this is unnecessary if every element of A is its own opposite.

3. Nowhere-zero *k*-flows

Face Colorings and Flows

Consider a directed graph G which is 2-cell-embedded in an orientable surface S (see **[56]** and **ST1**, Chapter 2 for definitions). Assume that the embedding is *face-k-colorable*—that is, the faces of the embedding can be colored with k colors in such a way that each edge belongs to the boundary of two faces with different colors. Consider the colors as the elements of an additive group of order k—say, $\mathbf{Z}_k$. For each edge e, let $r(e)$ be the color of the face bounded by e on its right, and let $l(e)$ be the color of the face bounded by e on its left. Using remarks (*i*) and (*iii*) of Section 2, we can easily check that the mapping $r - l$ from $E(G)$ to $\mathbf{Z}_k$ is a $\mathbf{Z}_k$-flow of G. The face-coloring property is equivalent to the fact that this flow is nowhere-zero. We can formulate this result as follows:

Theorem 3.1. *If a graph has a face-k-colorable 2-cell embedding in some orientable surface, then it has a nowhere-zero* $\mathbf{Z}_k$*-flow.* ||

More can be said for plane embeddings (see **[49]**):

Theorem 3.2. *A plane graph is face-k-colorable if and only if it has a nowhere-zero* $\mathbf{Z}_k$*-flow.*

Sketch of proof. Each flow corresponds to a potential difference in the dual graph (this is the duality of flows and tensions for plane graphs mentioned in Section 1—see [**36**, Chapter 7]). This means that in the graph, each flow can be obtained by assigning a value to each face, and then assigning to each edge the difference between the value of the face on the right and the value of the face on the left. In particular, each nowhere-zero $\mathbf{Z}_k$-flow can be obtained from a face-k-coloring by the process used to prove Theorem 3.1. ||

Some Equivalence Results

Theorem 3.2 led Tutte to study nowhere-zero $\mathbf{Z}_k$-flows for arbitrary graphs. In particular he obtained two equivalence results that we now present briefly.

It is clear that in Theorems 3.1 and 3.2, the group $\mathbf{Z}_k$ can be replaced by any other additive group of the same order. This is a general phenomenon which is a consequence of Theorem 3.4 below.

For a graph G, and $F \subseteq E(G)$, we denote by $r(F)$ the maximum number of edges in a forest of G contained in F. We shall use the following lemma, which is an immediate extension of a classical result:

Lemma 3.3. *Let G be a connected directed graph, and let T be a spanning tree of G. Let A be an additive group, and let c be any mapping from $E(G) - E(T)$ to A. Then there exists exactly one A-flow φ of G such that, for each edge e of G not in T, $\varphi(e) = c(e)$.* ||

The following result is due to Tutte [**49**]:

Theorem 3.4. *Let A be a finite additive group of order λ, and let G be a directed graph. Then the number of nowhere-zero A-flows of G is*

$$F(G, \lambda) = \sum_{F \subseteq E(G)} (-1)^{|E(G) - F|} \lambda^{|F| - r(F)}.$$

Proof. It follows from Lemma 3.3 that, for every $F \subseteq E(G)$, $\lambda^{|F| - r(F)}$ is the number of A-flows of the subgraph $(V(G), F)$ of G. Equivalently, $\lambda^{|F| - r(F)}$ is the number of A-flows of G whose support is contained in F. The result now follows by the inclusion–exclusion principle. ||

The polynomial $F(G, \lambda)$ is called the **flow polynomial** of G. It is, in a sense, dual to the classical chromatic polynomial, and can be evaluated similarly by a deletion–contraction process (see [**53**]).

Another important equivalence result was obtained by Tutte [**49**]: let k be an integer, $k \geqslant 2$, and let G be a directed graph. A **nowhere-zero *k*-flow of *G*** is a $\mathbf{Z}$-flow φ of G such that $0 < |\varphi(e)| < k$ for each e in $E(G)$.

Theorem 3.5. *A directed graph G has a nowhere-zero k-flow if and only if it has a nowhere-zero $\mathbf{Z}_k$-flow.*

Sketch of proof. If we replace each edge-value of a nowhere-zero k-flow of G by the corresponding value of $\mathbf{Z}_k$, we obtain a nowhere-zero $\mathbf{Z}_k$-flow. Conversely, if we replace each edge-value of a nowhere-zero $\mathbf{Z}_k$-flow by the corresponding integer in $[1, k - 1]$, we obtain a mapping f from $E(G)$ to $\{1, 2, \ldots, k - 1\}$ which satisfies the following property:

for each vertex v of G,

$$\sum_{e \in \omega^+(\{v\})} f(e) - \sum_{e \in \omega^-(\{v\})} f(e) \equiv 0 \pmod{k}.$$

We consider a new graph G' obtained from G by adding a new vertex joined to each other vertex by a new edge. The mapping f yields a $\mathbf{Z}$-flow f' of G' which takes its values in $\{1, \ldots, k - 1\}$ for the edges of G, and in $k\mathbf{Z}$ for the new edges. A result of Tutte on regular chain-groups [**50**, Proposition 6.3] then allows us to derive from f' a nowhere-zero k-flow of G. ||

Direct proofs of Theorem 3.5 can be found in [**8**] and [**58**].

In view of Theorems 3.4 and 3.5, we define (as in [**22**]) a graph G to be an $\boldsymbol{F_k}$ **graph** (for $k \geq 2$) if it satisfies the following equivalent properties:

(*i*) for some additive group A of order k, G has a nowhere-zero A-flow;

(*ii*) for every additive group A of order k, G has a nowhere-zero A-flow;

(*iii*) G has a nowhere-zero k-flow.

Note that, by (*iii*), if G is an F_k graph, then it is an F_l graph for each $l \geq k$.

Clearly a graph is an F_2 graph if and only if all of its vertices have even degree (by property (*i*), with $A = \mathbf{Z}_2$). The following simple result gives interesting examples for $k = 3$ and $k = 4$ (see [**33**], [**48**], [**49**]):

Theorem 3.6. *Let G be a trivalent graph. Then*

(*i*) *G is an F_3 graph if and only if it is bipartite*;

(*ii*) *G is an F_4 graph if and only if it is edge-3-colorable.*

Sketch of proof. Part (*i*) is proved by considering nowhere-zero $\mathbf{Z}_3$-flows as orientations for which every vertex is a source or a sink.

Part (*ii*) is proved by considering nowhere-zero $\mathbf{Z}_2{}^2$-flows as edge-colorings with three colors. ||

4. *k*-flow Conjectures and Theorems

The 4-flow Conjecture

By Theorem 3.2, the four-color theorem (see [**1**]) is equivalent to the result that every bridgeless planar graph is an F_4 graph. In [**52**] Tutte conjectured the following stronger property:

The 4-flow conjecture. *Every bridgeless graph with no subgraph contractible to the Petersen graph is an F_4 graph.*

This conjecture is discussed in [**42**]. In this direction, using the four-color theorem and matroid theory, Walton and Welsh [**54**] proved that every bridgeless graph with no subgraph contractible to the Kuratowski graph $K_{3,3}$ is an F_4 graph.

It is apparently not known whether the 4-flow conjecture is equivalent to its restriction to trivalent graphs. This restriction can be formulated as follows (see Theorem 3.6 (*ii*)):

The trivalent 4-flow conjecture. *Every bridgeless trivalent graph with no subgraph homeomorphic to the Petersen graph is edge-3-colorable.*

A first (small) step towards a proof of this conjecture is the result that bridgeless trivalent graphs with crossing-number 1 are edge-3-colorable (see **[23]**, **[13]**). Of course, the proofs rely on the four-color theorem.

The 5-flow Conjecture

Tutte also looked for an analogue of the four-color theorem for arbitrary graphs. In **[49]**, he proposed the following conjecture:

The 5-flow conjecture. *Every bridgeless graph is an F_5 graph.*

Since the Petersen graph is not an F_4 graph (Theorem 3.6 (*ii*)), this conjecture, if true, would be the best possible.

Tutte also proposed the weaker conjecture that there exists an integer $k \geq 5$ such that every bridgeless graph is an F_k graph. This was proved for $k = 8$ in 1975 independently by Kilpatrick **[30]** and Jaeger **[20]**, **[22]**, using essentially the same method. This result is now superseded by the 6-flow theorem of Seymour (see below). However, we shall present a full proof here, because it is fairly simple and uses auxiliary results which are interesting for their own sake. We shall need two lemmas; the first one is due to Kundu **[31]**:

Lemma 4.1. *Every 2k-edge-connected graph* ($k \geq 1$) *contains k pairwise edge-disjoint spanning trees.*

Proof. Tutte **[51]** and Nash-Williams **[35]** have proved that a graph G contains k pairwise edge-disjoint spanning trees if and only if, for each partition P of $V(G)$ into p blocks, the number $m(P)$ of edges of G joining different blocks is at least $k(p-1)$. This is clearly true if $p = 1$. If $p \geq 2$, $P = \{B_1, \ldots, B_p\}$, and G is $2k$-edge-connected, then

$$m(P) = \tfrac{1}{2} \sum_{i=1}^{p} |\omega(B_i)| \geq \tfrac{1}{2}p(2k) > k(p-1). \;\|$$

The proofs of the next lemma given in **[22]** and **[30]** rely on a formula of Edmonds **[10]** on the minimum number of independent sets of a matroid needed to cover the elements. We give here a simpler proof:

Lemma 4.2. *Every 3-edge-connected graph has three spanning trees with empty intersection.*

Proof. Consider a 3-edge-connected graph G. Replacing every edge of G by two parallel edges, we obtain a 6-edge-connected graph G'. By Lemma 4.1, G' has three pairwise edge-disjoint spanning trees. By identifying each of these trees with a tree of G, we obtain three spanning trees of G with empty intersection. ||

We now prove the following 8-flow theorem:

Theorem 4.3. *Every bridgeless graph is an F_8 graph.*

Proof. It is easily seen that it is sufficient to prove the result for a 3-edge-connected graph G (see Section 9 below). By Lemma 4.2, we can find spanning trees T_1, T_2, T_3 of G which have empty intersection. Using Lemma 3.3, we can obtain $\mathbf{Z}_2$-flows φ_1, φ_2, φ_3 of G such that $\varphi_i(e) = 1$ for all e in $E(G) - E(T_i)(i = 1, 2, 3)$. Then $(\varphi_1, \varphi_2, \varphi_3)$ defines in the obvious way a nowhere-zero $\mathbf{Z}_2{}^3$-flow of G.||

The 6-flow Theorem

We present here an outline of Seymour's proof of this result, but from a slightly different perspective.

Consider the following constructions for a graph G:

C_0: add an isolated vertex to G;

C_1: add an edge within one connected component of G;

C_2: add two edges joining two distinct connected components of G.

Let $\mathscr{C}$ be the class of graphs which can be obtained from the graph K_1 by a finite number of constructions of the form C_0, C_1 and/or C_2. The following result implies that every graph in $\mathscr{C}$ is an F_3 graph:

Theorem 4.4. *Let G be a graph in $\mathscr{C}$, considered with an arbitrary orientation. For each mapping c from $E(G)$ to $\mathbf{Z}_3$, there exists a $\mathbf{Z}_3$-flow φ of G such that $\varphi(e) \neq c(e)$ for each e in $E(G)$.*

Proof. We proceed by induction on $|E(G)|$. If $|E(G)| = 0$, there is nothing to prove. Suppose that G' is constructed from G using construction C_1 or C_2, and let c' be a mapping from $E(G')$ to $\mathbf{Z}_3$. Let μ be a $\mathbf{Z}_3$-flow of G' whose support is a cycle containing $E(G') - E(G)$. Since $|E(G') - E(G)| \leq 2$, we may use μ to obtain a $\mathbf{Z}_3$-flow μ' of G' such that $\mu'(e) \neq c'(e)$ for each e in $E(G') - E(G)$. (μ' is equal to μ, $-\mu$ or the zero flow.) By the induction hypothesis, there exists a $\mathbf{Z}_3$-flow φ of G such that $\varphi(e) \neq c'(e) - \mu'(e)$, for each e in $E(G)$. Then $\varphi' = \varphi + \mu'$ is a $\mathbf{Z}_3$-flow of G' such that $\varphi'(e) \neq c'(e)$, for each e in $E(G')$. ||

We now present Seymour's 6-flow theorem **[41]**:

Theorem 4.5. *Every bridgeless graph is an F_6 graph.*

Sketch of proof. It is sufficient to prove the result for a simple 3-connected graph G (see Section 9 below). Seymour showed that there exist vertex-disjoint cycles $C_1, \ldots, C_r$ of G such that the graph H obtained by contracting the edges of these cycles belongs to $\mathscr{C}$. By Theorem 4.4, H has a nowhere-zero $\mathbf{Z}_3$-flow, which can be extended to a $\mathbf{Z}_3$-flow φ_3 of

G with $E(G) - \bigcup_{i=1}^{r} C_i \subseteq \sigma(\varphi_3)$. Consider now a $\mathbf{Z}_2$-flow φ_2 of G with $\sigma(\varphi_2) = \bigcup_{i=1}^{r} C_i$. Then (φ_2, φ_3) defines a nowhere-zero $(\mathbf{Z}_2 \times \mathbf{Z}_3)$-flow of G. ||

Seymour **[41]** has also given a sketch of a proof of the following result, which yields an alternative proof of the 6-flow theorem:

Theorem 4.6. *Let G be a 3-connected trivalent simple graph. Then there exists a spanning tree T of G such that the contraction of the edges of $E(G) - T$ yields a graph which belongs to $\mathscr{C}$.* ||

In **[58]** Younger used Seymour's proof to obtain a polynomial-time algorithm for constructing a nowhere-zero 6-flow in any bridgeless graph. For planar graphs this algorithm can be specialized to yield a nowhere-zero 5-flow.

The 3-flow Conjecture

A theorem of Grötzsch **[17]** asserts that every loopless planar graph without triangles is vertex-3-colorable. By duality and Theorem 3.2, this can be reformulated as follows: every bridgeless planar graph without 3-cuts is an F_3 graph. This led Tutte to propose the following conjecture (see [**6**, unsolved problem 48]); it is easy to see that it would be sufficient to prove this conjecture for 4-edge-connected graphs.

The 3-flow conjecture. *Every bridgeless graph without 3-cuts is an F_3 graph.*

The following result appears in **[20]**, **[22]**:

Theorem 4.7. *Every bridgeless graph without 3-cuts is an F_4 graph.*

Proof. It is easy to show that it is sufficient to prove the result for 4-edge-connected graphs. By Lemma 4.1, any such graph G contains two edge-disjoint spanning trees T_1, T_2. By Lemma 3.3, there exists a $\mathbf{Z}_2$-flow $\varphi_i (i = 1, 2)$ of G such that $E(G) - E(T_i) \subseteq \sigma(\varphi_i)$. Then (φ_1, φ_2) is a nowhere-zero $\mathbf{Z}_2{}^2$-flow of G. ||

We propose the following conjecture:

The weak 3-flow conjecture. *There exists an integer k such that every k-edge-connected graph is an F_3 graph.*

A possible approach to this conjecture would be to enlarge the class $\mathscr{C}$ introduced above to a class $\mathscr{C}'$, by allowing new constructions which preserve the F_3 property (for instance, insertion of a new vertex into an edge, or identification of vertices). We might then ask whether there exists an integer k such that every k-edge-connected graph is in $\mathscr{C}'$.

A More General Conjecture

Let us call a graph **mod (2*p* + 1)-orientable** ($p \geq 1$) if it has an orientation such that the out-degree of each vertex is congruent (modulo $2p + 1$) to the in-degree. We denote by $U(\mathbf{Z}_{2p+1})$ the subset $\{\overline{1}, -\overline{1}\}$ of $\mathbf{Z}_{2p+1}$. We then obtain the following result:

Theorem 4.8. *For any graph G, and for any* $p \geq 1$, *the following properties are equivalent*:

(*i*) *G is mod* $(2p + 1)$*-orientable*;

(*ii*) *G has a* $U(\mathbf{Z}_{2p+1})$*-flow*;

(*iii*) *G has a* $(\mathbf{Z} \cap ([-p - 1, -p] \cup [p, p + 1]))$*-flow*;

(*iv*) *G has a* $(\mathbf{Q} \cap ([-1 - 1/p, -1] \cup [1, 1 + 1/p]))$*-flow.*

Sketch of proof. The equivalence of (*i*) and (*ii*) is immediate. The equivalence of (*iii*) and (*iv*) is a special case of a well-known property of flows. The equivalence of (*ii*) and (*iii*) is proved in two steps. The first step is that (*ii*) holds if and only if *G* has a $(\mathbf{Z}_{2p+1} \cap \{\overline{p}, \overline{p + 1}\})$-flow; this is easily proved by using an appropriate automorphism of $\mathbf{Z}_{2p+1}$. The equivalence between (*iii*) and the existence of a $(\mathbf{Z}_{2p+1} \cap \{\overline{p}, \overline{p + 1}\})$-flow of *G* is then proved by the same method as Theorem 3.5. ||

Note that *G* is mod 3-orientable if and only if it is an F_3 graph.

For every $p \geq 1$, let M_{2p+1} be the class of mod $(2p + 1)$-orientable graphs. It follows from the equivalence of (*i*) and (*iv*) of Theorem 4.8 that $M_{2p'+1} \subseteq M_{2p+1}$ for all $p' \geq p \geq 1$. It is not difficult to check that, for every $p \geq 1$, the complete graph K_{4p+2} belongs to $M_{2p+1} - M_{2p+3}$. Hence $M_{2p'+1} \subset M_{2p+1}$, for all $p' > p \geq 1$. It is then tempting to conjecture that the higher the edge-connectivity of a graph, the higher it is ranked in the hierarchy $M_3 \supset M_5 \supset \ldots \supset M_{2p+1} \supset \ldots$. To make this more precise, the following conjecture (whose name is justified in Section 5) is proposed in [**25**]:

The circular flow conjecture. *For all* $p \geq 1$, *every* $4p$*-edge-connected graph is mod* $(2p + 1)$*-orientable.*

Note that, for $p = 1$, this conjecture is equivalent to the 3-flow conjecture. Moreover, it is easy to see that for $p = 2$, the conjecture implies the 5-flow conjecture. Indeed, assume that every 8-edge-connected graph is mod 5-orientable. Consider a 3-edge-connected graph G and replace each edge of G by three parallel edges. The resulting graph G' is 9-edge-connected, and hence mod 5-orientable. It is then easy to convert a $U(\mathbf{Z}_5)$-flow of G' into a nowhere-zero $\mathbf{Z}_5$-flow of G. The extension to all bridgeless graphs is immediate (see Section 9).

Balanced Valuations

Like the four-color problem, the nowhere-zero flow problems presented in this section lend themselves to various interesting reformulations. We shall consider only one of these here.

A **balanced valuation** of a graph G is a mapping b from $V(G)$ to Q such that,

$$\text{for all } S \subseteq V(G),\ \left|\sum_{v \in S} b(v)\right| \leq |\omega_G(S)|.$$

The following result is proved in **[19]**:

Theorem 4.9. *Let p, q be integers with $1 \leq p < q$. Then a graph G has a $\mathbf{Z} \cap ([-q, -p] \cup [p, q])$-flow if and only if it has a balanced valuation of the form $\dfrac{q+p}{q-p}w$, where w is a mapping from $V(G)$ to $\mathbf{Z}$ such that, for all $v \in V(G)$, $w(v)$ and the degree of v have the same parity.* ||

For instance, if we apply Theorem 4.9 with $p = 1$ and $q = 3$ to trivalent graphs, and if we then use Theorem 3.6(*ii*), we obtain the following result of Bondy **[4]**: *a trivalent graph is edge-3-colorable if and only if it has a balanced valuation with values in* $\{-2, +2\}$. Similarly, it follows from Theorem 4.9 that *a 5-regular graph is an F_3 graph if and only if it has a balanced valuation with values in* $\{-3, +3\}$ (see **[5]**). More generally, all the results and problems discussed above can be reformulated in terms of balanced valuations. For instance, the 6-flow theorem asserts that every bridgeless trivalent graph has a balanced valuation with values in $\{-\frac{3}{2}, +\frac{3}{2}\}$, and the 5-flow conjecture asks whether $\frac{3}{2}$ can be replaced by $\frac{5}{3}$ in this statement.

5. The Double Cover Conjecture

A 2-cell embedding of a graph G on a surface is a **strong embedding** if each face-boundary is a cycle. For instance, every embedding of a planar 2-connected graph in the plane is a strong embedding.

A **cycle double cover** of a graph G is a family of cycles of G such that every edge appears in exactly two cycles of this family. For instance, the family of face-boundaries of a strong embedding of G is a cycle double cover.

The following two conjectures appear in the literature (see **[18]**, **[32]**, **[39]**, **[45]** and **[57]**):

The strong embedding conjecture. *Every 2-connected graph has a strong embedding on some surface.*

The double cover conjecture. *Every bridgeless graph has a cycle double cover.*

The double cover conjecture is easily seen to be equivalent to its restriction to 2-connected graphs. Hence, *the strong embedding conjecture implies the double cover conjecture*. The first conjecture appears as a topological motivation for the second one. Both problems are reviewed in some detail in **[26]**. Our concern here is only with the double cover conjecture which, as we shall see, can be viewed as a nowhere-zero flow problem.

A cycle double cover is said to be ***k*-colorable** ($k \geq 2$) if we can color its cycles with k colors in such a way that each edge appears on two cycles of different colors. For instance, a cycle double cover consisting of k cycles is k-colorable. The family of face-boundaries of a face-k-colorable strong embedding of G is a k-colorable cycle double cover of G.

Let D_k be the subset of $\mathbf{Z}_2{}^k$ consisting of those elements containing exactly two 1s, and let $\varphi = (\varphi_1, \ldots, \varphi_k)$ be a D_k-flow of G. For each $i \in \{1, \ldots, k\}$, choose a partition of $\sigma(\varphi_i)$ into cycles of G. The union of these k partitions is a k-colorable cycle double cover of G. Conversely, it is easily seen that every k-colorable cycle double cover of G can be obtained in this way from some D_k-flow of G. Thus we obtain the following reformulation of the double cover conjecture:

The double cover conjecture (second version). *For every bridgeless graph G, there exists an integer $k \geq 2$ such that G has a D_k-flow.*

The following stronger conjecture appears in **[8]** and **[37]**:

The 5-colorable double cover conjecture. *Every bridgeless graph has a D_5-flow.*

A cycle double cover is said to be **orientable** if one can orient each of its cycles into a directed cycle in such a way that each edge appears once in each direction in the resulting family of directed cycles. For instance, the family of face-boundaries of a strong embedding of G in some orientable surface is an orientable cycle double cover.

Let $\vec{D}_k$ be the subset of $\mathbf{Z}^k$ ($k \geq 2$) consisting of those elements ($z_1, \ldots, z_k$) such that exactly one z_i is 1, exactly one z_i is -1, and all the other z_i are 0. It is easy to see that *G has an orientable k-colorable cycle double cover if and only if it has a* $\vec{D}_k$*-flow.*

Oriented versions of the previous conjectures can be stated as follows:

The orientable double cover conjecture. *For every bridgeless graph G, There exists an integer* $k \geq 2$ *such that G has a* $\vec{D}_k$*-flow.*

The 5-colorable orientable double cover conjecture. *Every bridgeless graph has a* $\vec{D}_5$*-flow.*

These oriented versions are related to nowhere-zero k-flow problems. Indeed, it is clear that the proof of Theorem 3.1 can be immediately adapted to yield the following:

Theorem 5.1. *If a graph has a* $\vec{D}_k$*-flow* ($k \geq 2$), *then it is an* F_k *graph.* ||

An interesting consequence of this result is that the 5-colorable orientable double cover conjecture implies the 5-flow conjecture. If we drop the orientability of cycle double covers, we can prove only the following result:

Theorem 5.2. *If a graph has a* D_k*-flow* ($k \geq 2$), *then it is an* F_p *graph for* $p = 2^{\lceil \log_2 k \rceil}$.

Proof. Let $\varphi = (\varphi_1, \ldots, \varphi_k)$ be a D_k-flow of G. Let $r = \lceil \log_2 k \rceil$, and let f be a one-to-one mapping from $\{1, \ldots, k\}$ into $\mathbf{Z}_2{}^r$. For each i in $\{1, \ldots, k\}$, let φ_i' be the $\mathbf{Z}_2{}^r$-flow of G which takes the value $f(i)$ on $\sigma(\varphi_i)$ and the value 0 on $E(G) - \sigma(\varphi_i)$. It is easy to check that $\sum_{i=1}^{k} \varphi_i'$ is a nowhere-zero $\mathbf{Z}_2{}^r$-flow of G. ||

Theorems 5.1 and 5.2 have the following converses for small values of k; Theorem 5.4 was proved for trivalent graphs by Tutte [**48**]:

Theorem 5.3. *The following statements are equivalent for a graph G:*

(*i*) *G is an* F_2 *graph;*

(*ii*) *G has a* $\vec{D}_2$*-flow;*

(*iii*) *G has a* D_2*-flow.*

Proof. This is immediate. ||

Theorem 5.4. *The following statements are equivalent for a graph G:*

(*i*) *G is an* F_3 *graph;*

(*ii*) *G has a* $\vec{D}_3$*-flow.*

Proof. This is the case $p = 1$ of [**25**, Proposition 2]. ||

Theorem 5.5. *The following statements are equivalent for a graph G:*

(*i*) *G is an* F_4 *graph;*

(*ii*) *G has a* D_3*-flow;*

(*iii*) *G has a* $\vec{D}_4$*-flow;*

(*iv*) *G has a* D_4*-flow.*

Proof. (*i*) ⇒ (*ii*). If (φ_1, φ_2) is a nowhere-zero $\mathbf{Z}_2{}^2$-flow of G, then $(\varphi_1, \varphi_2, \varphi_1 + \varphi_2)$ is a D_3-flow of G.

(*ii*) ⇒ (*iii*) (see [**48**]). Let $(\varphi_1, \varphi_2, \varphi_3)$ be a D_3-flow of G. Fir $i = 1, 2, 3$, let φ_i' be a **Z**-flow of G which takes the value 1 or -1 on $\sigma(\varphi_i)$ and the value 0 on $E(G) - \sigma(\varphi_i)$. Let

$$\psi_1 = \tfrac{1}{2}(\varphi_1' - \varphi_2' - \varphi_3'),\ \psi_2 = \tfrac{1}{2}(-\varphi_1' + \varphi_2' - \varphi_3'),$$
$$\psi_3 = \tfrac{1}{2}(-\varphi_1' - \varphi_2' + \varphi_3') \text{ and } \psi_4 = \tfrac{1}{2}(\varphi_1' + \varphi_2' + \varphi_3').$$

Then $(\psi_1, \psi_2, \psi_3, \psi_4)$ is a $\vec{D}_4$-flow of G.

(*iii*) ⇒ (*iv*). This is immediate.

(*iv*) ⇒ (*i*). This is Theorem 5.2 for $k = 4$. ||

The property of being mod $(2p + 1)$-orientable (see Section 4) can also be studied in terms of cycle double covers of a special kind. Consider a vector $\mathbf{x} = (x_1, \ldots, x_k)$, where the x_i belong to $\mathbf{Z}_2$ or $\mathbf{Z}$. Two of the coordinates x_i and x_j are said to be **cyclically consecutive** if $|j - i| = 1$ or $k - 1$. Let C_k be the subset of D_k consisting of those elements whose two non-zero components are cyclically consecutive. Similarly, let $\vec{C}_k$ be the subset of $\vec{D}_k$ consisting of those elements whose two non-zero components are cyclically consecutive. The following result was proved in [**25**]; Theorem 5.4 is the special case $p = 1$ of this result:

Theorem 5.6. *A graph is mod* $(2p + 1)$*-orientable* $(p \geq 1)$ *if and only if it has a* $\vec{C}_{2p+1}$*-flow.* ||

It follows from Theorem 5.6 that the circular flow conjecture of Section 4 can be reformulated as follows:

The circular flow conjecture (second version). *For all* $p \geq 1$, *every* $4p$*-edge-connected graph has a* $\vec{C}_{2p+1}$*-flow.*

In support of this conjecture, the following result is proved in [**25**]:

Theorem 5.7. *For all* $p \geq 1$, *every* $4p$*-edge-connected graph has a* C_{2p+1}*-flow.* ||

For $p = 1$, this is essentially the nowhere-zero-4-flow result of Theorem 4.7. The proof of the general case also relies on Lemma 4.1.

6. Fulkerson's Conjecture

We present here a conjecture of Fulkerson [15], which deals with 1-factors of trivalent graphs and can be viewed as an edge-coloring problem. As we shall see, it can be reformulated as a nowhere-zero flow problem.

Fulkerson's conjecture. *In every bridgeless trivalent graph, there exists a family of six 1-factors such that each edge appears in exactly two of them.*

A trivalent graph which satisfies this condition is said to have the **Fulkerson property**. The following result is a characterization of this property in terms of flows. We denote by Y the subset of $\mathbf{Z}_2{}^6$ consisting of those elements containing exactly four 1s.

Theorem 6.1. *Let G be a trivalent graph. Then G has the Fulkerson property if and only if G has a Y-flow.*

Proof. Let $(\varphi_1, \varphi_2, \varphi_3, \varphi_4, \varphi_5, \varphi_6)$ be a Y-flow of G, and let v be a vertex of G. Then

$$\sum_{i=1}^{6} |\sigma(\varphi_i) \cap \omega(\{v\})| = \sum_{e \in \omega(\{v\})} \sum_{i=1}^{6} |\sigma(\varphi_i) \cap \{e\}| = 4|\omega(\{v\})| = 12.$$

Since $|\sigma(\varphi_i) \cap \omega(\{v\})| \in \{0, 2\}$ for $i = 1, \ldots, 6$, we thus have $|\sigma(\varphi_i) \cap \omega(\{v\})| = 2$. Hence, $M_i = E(G) - \sigma(\varphi)$ is a 1-factor of G, for $i = 1, \ldots, 6$, and each edge of G belongs to exactly two of the sets M_i.

Conversely, if G has the Fulkerson property, then it is easy to construct a Y-flow of G, since to every 1-factor M there corresponds a $\mathbf{Z}_2$-flow φ with $\sigma(\varphi) = E(G) - M$. ||

It is not difficult to show that, if every trivalent bridgeless graph has a Y-flow, then the same property holds for all bridgeless graphs (see Section 9). Hence we can reformulate Fulkerson's conjecture as follows:

Fulkerson's conjecture (second version). *Every bridgeless graph has a Y-flow.*

This conjecture can also be viewed as part of a wider problem. For $p \geq 1$, let Y_p be the subset of $\mathbf{Z}_2{}^{3p}$ consisting of those elements with exactly $2p$ 1s. Thus Y_1 is the set D_3 introduced in Section 5, and Y_2 is the set Y defined above. It follows from work of Seymour [40], and from an easy extension of the above discussion, that for each bridgeless graph G, the set $P(G)$ of positive integers p such that G has a Y_p-flow contains all sufficiently large integers, or all sufficiently large even integers. Moreover, the first situation occurs if G is trivalent and has no subgraph homeomorphic to the Petersen graph—this should be compared with the trivalent 4-flow conjecture. Any upper bound for min $P(G)$, valid for all bridgeless graphs G, would constitute important progress.

7. A Unifying Conjecture

So far we have encountered a number of conjectures in which B is a subset of some additive group A, such that $0 \notin B$ and $B = -B$, and every bridgeless graph is conjectured to have a B-flow. Let us call this the ***B*-flow conjecture**.

We present here a conjecture which implies all reasonable B-flow conjectures when B is a subset of $\mathbf{Z}_2{}^k$, for some integer k. This conjecture appears in **[24]** in the context of binary spaces, but we reformulate it in purely graph-theoretical terms.

In this section we consider $\mathbf{Z}_2$ as a field—that is, we write $\mathbf{Z}_2$ for $GF(2)$. It is well known that the set of $\mathbf{Z}_2$-flows of a graph G forms a vector space over $\mathbf{Z}_2$. We denote this vector space by $\mathcal{F}(G)$ and its dimension by $\mu(G)$. Recall that a *subdivision* of a graph G is any graph which can be obtained from G by inserting new vertices of degree 2 into the edges.

Let G_1 and G_2 be two graphs. We write $G_1 \leq G_2$ if there exists a subdivision G_1' of G_1 with the following property: there exists a bijective mapping β from $E(G_2)$ to $E(G_1')$ such that, for each $\mathbf{Z}_2$-flow φ of G_1', $\varphi \circ \beta$ is a $\mathbf{Z}_2$-flow of G_2. We write $G_1 \simeq G_2$ if $G_1 \leq G_2$ and $G_2 \leq G_1$.

For instance, denote by K_2^3 the graph consisting of two vertices and three parallel edges e_1, e_2, e_3 joining these two vertices, and let G be an edge-3-colorable trivalent graph.

We show that $K_2^3 \leq G$. For this, consider an edge-3-coloring of G, and let $M_i (i = 1, 2, 3)$ be the set of edges of G of the ith color. For $i = 1, 2, 3$, replace e_i by a path P_i of length $|M_i|$, and let H be the resulting subdivision of K_2^3. There exists a bijective mapping β from $E(G)$ to $E(H)$ such that $E(P_i) = \beta(M_i) (i = 1, 2, 3)$. Then, for each $\mathbf{Z}_2$-flow φ of H which is not identically zero, $\varphi \circ \beta$ is a $\mathbf{Z}_2$-flow of G whose support is a set of bicolored cycles. Hence $K_2^3 \leq G$.

The following result is easy to prove:

Theorem 7.1. *The relation $\leq$ is a quasi-order—that is, it is reflexive and transitive. If $G_1 \leq G_2$, then $|E(G_1)| \leq |E(G_2)|$, and $\mu(G_1) \leq \mu(G_2)$; moreover, equality holds in both inequalities if and only if $G_1 \simeq G_2$.* $\|$

It follows that $G_1 \simeq G_2$ if and only if $\mathcal{F}(G_1)$ and $\mathcal{F}(G_2)$ are isomorphic in a strong sense—that is, if and only if the cycle matroids of G_1 and G_2 are isomorphic (for definitions, see Chapter 3 or [**55**]).

Let $\mathscr{C}$ be a class of graphs. A graph G is said to be **$\mathscr{C}$-minimal** if G belongs to $\mathscr{C}$, and $G' \simeq G$ whenever G' belongs to $\mathscr{C}$ and $G' \leq G$.

Theorem 7.2. *For each graph G in $\mathscr{C}$, there exists a $\mathscr{C}$-minimal graph G_0 such that $G_0 \leq G$.*

Proof. This is an immediate consequence of Theorem 7.1. ||

Now let k be a positive integer, and let B be a subset of $\mathbf{Z}_2{}^k - \{0\}$. The following result motivates our study of the relation $\leq$:

Theorem 7.3. *If G_1 and G_2 are two graphs with $G_1 \leq G_2$, and if G_1 has a B-flow, then G_2 also has a B-flow.*

Proof. Let G_1' be a subdivision of G_1, and let β be a bijective mapping from $E(G_2)$ to $E(G_1')$ such that, for each $\mathbf{Z}_2$-flow φ of G_1', $\varphi \circ \beta$ is a $\mathbf{Z}_2$-flow of G_2. If G_1 has a B-flow, G_1' also has a B-flow $(\varphi_1, \ldots, \varphi_k)$. Then $(\varphi_1 \circ \beta, \ldots, \varphi_k \circ \beta)$ is a B-flow of G_2. ||

From now on, $\mathscr{C}$ denotes the class of bridgeless graphs. It follows from Theorems 7.2 and 7.3 that, for any $B \subseteq \mathbf{Z}_2{}^k - \{0\}$, the B-flow conjecture is equivalent to its restriction to $\mathscr{C}$-minimal graphs.

The only $\mathscr{C}$-minimal graphs known so far (up to equivalence under $\simeq$) are the graph L consisting of one vertex and one loop at this vertex, the graph K_2^3 defined above and the Petersen graph P. Moreover, it is easy to show that G is an F_2 graph if and only if $L \leq G$, and that G is an F_4 graph if and only if $L \leq G$ or $K_2^3 \leq G$.

In [**24**] we conjectured that L, K_2^3 and P are the only $\mathscr{C}$-minimal graphs, up to equivalence under $\simeq$. This conjecture essentially says that, for $B \subseteq \mathbf{Z}_2{}^k - \{0\}$ $(k \geq 1)$, the B-flow conjecture is true if and only if it holds for L, K_2^3 and P. Thus, for instance, our conjecture implies the 5-colorable double cover conjecture and Fulkerson's conjecture. We now reformulate our conjecture as a B-flow conjecture.

Let $\{\varphi_1, \varphi_2, \ldots, \varphi_6\}$ be a basis of the vector space $\mathscr{F}(P)$ of $\mathbf{Z}_2$-flows of the Petersen graph P, and let X be the subset of $\mathbf{Z}_2{}^6$ consisting of the fifteen elements of the form $(\varphi_1(e), \varphi_2(e), \ldots, \varphi_6(e))$, for $e \in E(P)$. We then have the following result:

Theorem 7.4. *The following properties are equivalent for a graph G:*

(*i*) *G has an X-flow;*

(*ii*) *$L \leq G$ or $K_2^3 \leq G$ or $P \leq G$.* ||

It follows that the property for a graph G to have an X-flow is independent of the choice of the basis $\{\varphi_1, \varphi_2, \ldots, \varphi_6\}$. We therefore say that G has a **Petersen flow** if G has an X-flow.

It is now clear that our conjecture about $\mathscr{C}$-minimal graphs can be reformulated as follows:

The Petersen flow conjecture. *Every bridgeless graph has a Petersen flow.*

The Petersen flow conjecture can be restricted to trivalent graphs (see Section 9). Moreover, it is easy to see that a trivalent graph G satisfies the Petersen flow conjecture if and only if we can color its edges, using the edges of the Petersen graph P as colors, in such a way that every triple of mutually incident edges of G is colored as a similar triple of P. Another simple formulation in terms of edge-5-colorings was recently obtained in [**27**].

A similar approach via a quasi-order relation for arbitrary B-flow problems is also undertaken in [**27**]. The quasi-order description and its properties are more complicated in this general case, and are not presented here.

8. Special Results

Most of the conjectures reviewed above are true for planar graphs and for F_4 graphs. It is therefore natural to try to prove them for classes of graphs which are either 'nearly planar' or 'nearly F_4'. This direction of research offers a variety of open problems, which should be more tractable than the general conjectures.

For graphs which are 'nearly planar', we could try, for instance, to prove some of the conjectures for graphs of small orientable or non-orientable genus. The first results we know in this direction were obtained by Steinberg, who proved the 3-flow conjecture [**43**] and the 5-flow conjecture [**44**] for graphs embeddable in the projective plane. More recently, the 5-flow conjecture has been proved independently by Möller *et al.* [**34**] and Fouquet [**14**] for graphs of orientable genus at most 2 and graphs of non-orientable genus at most 4.

We now present some results concerning graphs which are 'nearly F_4'. We define a graph G to be a **nearly-F_4 graph** if it is possible to add a new edge to G in order to obtain an F_4 graph. A graph G is a **deletion-F_4 graph** if it is possible to delete an edge of G in order to obtain and F_4 graph.

Examples of nearly-F_4 graphs are the following (see [**21**] for the first three classes):

(*i*) *F_4 graphs*: adding a loop, or an edge parallel to an existing edge, to an F_4 graph yields an F_4 graph.

(*ii*) *graphs with a Hamiltonian path*: by adding an edge, it is possible to obtain a graph with a Hamiltonian cycle—such a graph is easily seen to be an F_4 graph.

(*iii*) *trivalent graphs with a 2-factor having* 0 *or* 2 *odd components*: if the 2-factor has no odd cycles, then the graph is an F_4 graph (Theorem 3.6(*ii*)), whereas if the 2-factor has two odd cycles, then its edges can be colored with two colors (0, 1) and (1, 0) in such a way that each vertex, with the exception of two vertices v and v', is bicolored; if we now join v and v' by a new edge, and color the edges yet uncolored with (1, 1), then the resulting coloring of the edges defines a nowhere-zero $\mathbf{Z}_2{}^2$-flow.

(*iv*) *deletion-F_4 graphs*: if the deletion of an edge e yields an F_4 graph, adding a new edge parallel to e clearly gives an F_4 graph.

We now discuss some results on nearly-F_4 graphs, starting with the 5-flow conjecture (see [**21**]):

Theorem 8.1. *Every bridgeless nearly-F_4 graph is an F_5 graph.*

Sketch of proof. We must prove that if G is an F_4 graph and if $G - e$ is bridgeless for some $e \in E(G)$, then $G - e$ is an F_5 graph. It is possible to choose an orientation of G which has a $(\mathbf{Z} \cap \{1, 2, 3\})$-flow φ. It is easy to see that we may also assume that $\varphi(e) = 1$.

If $\omega^+(S) = \{e\}$ for some $S \subseteq V(G)$, then

$$\sum_{e' \in \omega^-(S)} \varphi(e') = \varphi(e) = 1,$$

and hence $|\omega^-(S)| = 1$. But this contradicts the hypothesis that $G - e$ is bridgeless. It follows that there exists in $G - e$ a directed path from the initial end of e to its terminal end. Hence there exists a $\mathbf{Z}$-flow μ of G such that $\mu(e) = -1$ and $\mu(e') \in \{0, 1\}$, for all e' in $E(G) - \{e\}$. It follows that $\varphi + \mu$ is a $\mathbf{Z}$-flow of G which takes the value 0 on e and values in $\{1, 4\}$ elsewhere, and we may consider $\varphi + \mu$ as a nowhere-zero 5-flow of $G - e$. ||

We next discuss some special cases of the double cover conjecture. The following result is due to Celmins [**8**]:

Theorem 8.2. *Every trivalent* 3*-edge-connected deletion-F_4 graph has a D_5-flow.*

Proof. Let G be a trivalent 3-edge-connected graph, and let $e \in E(G)$ be such that $G - e$ is an F_4 graph. It follows from the 3-edge-connectivity of G that there exists a cycle C of $G - e$ which contains both ends of e.

Let φ_0 be the $\mathbf{Z}_2$-flow of $G - e$ with $\sigma(\varphi_0) = C$, and let $(\varphi_1, \varphi_2, \varphi_3)$ be a D_3-flow of $G - e$. Then it is easy to check that $(\varphi_0, \varphi_0 + \varphi_1, \varphi_0 + \varphi_2, \varphi_0 + \varphi_3)$ is a D_4-flow of $G - e$.

The two ends of e define a partition of C into two paths P' and P''. Let φ_0' be the $\mathbf{Z}_2$-flow of G with support $P' \cup \{e\}$, and let φ_0'' be the $\mathbf{Z}_2$-flow of G with support $P'' \cup \{e\}$. Then $(\varphi_0', \varphi_0'', \varphi_0 + \varphi_1, \varphi_0 + \varphi_2, \varphi_0 + \varphi_3)$ is a D_5-flow of G. ||

The following result is due to Tarsi [**47**]; its proof relies on an ingenious construction:

Theorem 8.3. *Every bridgeless graph with a Hamiltonian path has a* D_6-*flow.* ||

9. Reductions

We consider here only 'B-flow conjectures', in the sense of Section 7—that is, conjectures concerning the class of all bridgeless graphs. The approach presented in this section can, of course, be used for any kind of nowhere-zero flow problem. For instance, the proof of Steinberg's 3-flow theorem for graphs embedded in the projective plane is a good example of the use of reduction techniques.

We start by presenting the reduction of the B-flow conjecture to trivalent 3-edge-connected graphs. In this section, A denotes an additive group and B is any non-empty subset of A with $0 \notin B$, $B = -B$. Recall that the B-flow conjecture states that every bridgeless graph has a B-flow. We shall assume that the B-flow conjecture is true for the graph K_2^3 defined in Section 7; equivalently, there exist b_1, b_2, b_3 in B such that $b_1 + b_2 + b_3 = 0$.

The following result is a generalized version of [**41**, Proposition 2.1], and is proved in a similar way:

Theorem 9.1. *Every minimal counter-example to the B-flow conjecture is a simple trivalent* 3-*edge-connected graph.*

Proof. A minimal counter-example G is clearly a loopless 2-edge-connected graph, which is not an F_2 graph, and has at least 3 edges. If $\{e_1, e_2\}$ is a 2-cut of G, we can contract e_1 and obtain a 2-edge-connected graph which, by the minimality of G, has a B-flow. This easily yields a B-flow of G, and we have a contradiction. Hence G is 3-edge-connected, and in particular has no vertices of degree smaller than 3.

We now assume that G has a vertex v of degree greater than 3. Then, by a result of Fleischner [**11**], we can find two edges e_1 and e_2 incident to v such that we obtain a bridgeless graph by deleting e_1 and e_2 and adding a new edge joining the ends of e_1, e_2 distinct from v. By the minimality of G, this graph has a B-flow, which easily gives a B-flow of G; again, we get a contradiction.

We conclude that G is trivalent and 3-edge-connected; G must be simple because otherwise G is isomorphic to K_2^3, which is not a counter-example. ||

Note that, by a well-known result, '3-edge-connected' can be replaced by '3-connected' in Theorem 9.1.

We now assume that the B-flow conjecture is true for all F_4 graphs. Hence, by Theorem 3.6(*ii*), every minimal counter-example to the B-flow conjecture is a simple trivalent 3-edge-connected graph which is not edge-3-colorable.

We call a 3-cut of a trivalent graph G *trivial* if it is of the form $\omega(\{v\})$ for some vertex v of G. A **snark** (see [**9**] or [**13**]) is a trivalent, 3-edge-connected, non-edge-3-colorable graph in which every 3-cut is trivial. We now present a condition under which the B-flow conjecture can be reduced to snarks.

Let G be a trivalent graph, and let v be a vertex of G. We choose an orientation of G such that the three edges e_1, e_2, e_3 incident to v have initial end v. We say that v is ***B*-specifiable** if, for all b_1, b_2, b_3 in B such that $b_1 + b_2 + b_3 = 0$, there exists a B-flow φ of G such that $\varphi(e_i) = b_i$, for $i = 1, 2, 3$. We say that B has the **vertex-specification property** if, for every trivalent graph G which has a B-flow, every vertex of G is B-specifiable.

Theorem 9.2. *If B has the vertex-specification property, then every minimal counter-example to the B-flow conjecture is a snark.*

Sketch of proof. Let G be a minimal counter-example, and assume that G has a non-trivial 3-cut $\omega(S)$, so that $|S| \geq 2$ and $|V(G) - S| \geq 2$. By identifying the vertices of S to a single vertex, we obtain a bridgeless trivalent graph G'. Similarly, by identifying the vertices of $V(G) - S$ to a single vertex, we obtain a bridgeless trivalent graph G''. By the definition of G, the graphs G' and G'' have B-flows. The vertex-specification property gives B-flows of G' and G'' which can be 'pieced together' into a B-flow of G. This gives the required contradiction. ||

In most of the interesting cases, the vertex-specification property can easily be proved using symmetry considerations. This is true, for instance, when B is D_k, $\vec{D}_k$ for $k \geq 3$ (see Section 5), or Y_p for $p \geq 1$ (see Section 6). For $B = \mathbf{Z}_5 - \{0\}$, the result is non-trivial. It is proved in [**8**], using the contraction–deletion process associated with the computation of the flow polynomial to obtain a stronger enumerative result.

An even more difficult result is proved in **[8]**:

Theorem 9.3. *Every minimal counter-example to the* 5*-flow conjecture is a cyclically* 5*-edge-connected snark of girth at least* 7.

Outline of proof. The proof that every minimal counter-example is cyclically 5-edge-connected uses sophisticated enumerative methods analogous to the Birkhoff–Lewis reduction of the 4-ring for the four-color problem **[3]**. To see how the proof of the girth property works, consider, for instance, a cycle C of length 6 in a minimal counter-example G. Delete three edges which form a perfect matching of C. Since G is cyclically 5-edge-connected, the resulting graph G' is bridgeless. It is then easy to combine a nowhere-zero $\mathbf{Z}_5$-flow of G' with a $\mathbf{Z}_5$-flow of G whose support is C to obtain a nowhere-zero $\mathbf{Z}_5$-flow of G. This gives the required contradiction. ||

An analogous result holds for counter-examples to the 5-flow conjecture which are minimal in the class of graphs embedded in a given surface S. Such a graph G is simple, 3-edge-connected, trivalent, of girth at least 6, and has no face-boundary of length less than 7 (see **[14]**, **[34]**). It follows easily from Euler's formula that $|V(G)| \leqslant -14k(S)$, where $k(S)$ is the Euler characteristic of S. Hence the 5-flow conjecture for graphs embedded in a given surface has been reduced to a finite number of cases.

Finally, a similar result was recently obtained by Goddyn **[16]**, for the double cover conjecture:

Theorem 9.4. *Every minimal counter-example to the double cover conjecture has girth at least* 7. ||

Theorems 9.3 and 9.4 are interesting in view of the fact that no snark with girth at least 7 is known. It is conjectured in **[28]** that such snarks do not exist.

10. Conclusion

In this chapter we have presented a brief survey of a rich class of problems which are rather strongly interrelated—the class of nowhere-zero flow problems. The basic problems appear to be difficult, but a number of reduction results and special results have been obtained, and more can be done in this direction.

To conclude, we mention some relationships between nowhere-zero flow problems and other areas of research.

The existence of nowhere-zero flows provides tools for obtaining cycle covers of graphs (that is, families of cycles covering all edges) of short total length. For instance, it is proved in **[2]** and **[46]** that every bridgeless graph G has a cycle cover of total length at most $\frac{5}{3}|E(G)|$. It is shown in [**38**] how some improvements could be derived from the validity of various B-flow conjectures.

Recently, Bouchet has initiated the study of nowhere-zero k-flows in bidirected graphs, proposing a '6-flow conjecture' and proving a '216-flow theorem' [**7**]. This work has been pursued by different authors ([**12**], [**29**], [**59**]), and the universal bound of 216 has been reduced to 30.

Finally, nowhere-zero flow problems have natural extensions to matroids and chain groups. For instance, the double cover conjecture can be viewed as a problem on binary matroids, and the 5-flow conjecture is a special case of the study of the critical exponent of matroids representable over $GF(5)$ (see Chapter 3 and [**55**]). It is likely that a significant advance on nowhere-zero flow problems would have interesting repercussions in the study of more general matroid problems.

References

1. K. Appel, W. Haken and J. Koch, Every planar map is four colorable, *Illinois J. Math.* **21** (1977), 429–567; *MR***58**#27598a/b.
2. J.-C. Bermond, B. Jackson and F. Jaeger, Shortest coverings of graphs with cycles, *J. Combinatorial Theory (B)* **35** (1983), 297–308; *MR***86a**:05078.
3. G. D. Birkhoff and D. C. Lewis, Chromatic polynomials, *Trans. Amer. Math. Soc.* **60** (1946), 355–451; *MR***8**–284f.
4. J. A. Bondy, Balanced colourings and the four colour conjecture, *Proc. Amer. Math. Soc.* **33** (1972), 241–244; *MR***45**#3246.
5. J. A. Bondy, Balanced colourings and graph orientation, *Proc. Sixth Southeastern Conference on Combinatorics, Graph Theory, and Computing* (ed. F. Hoffman *et al.*), *Congressus Numerantium XIV*, Utilitas Math., Winnipeg, 1975, pp. 109–114; *MR***53**#13014.
6. J. A. Bondy and U. S. R. Murty, *Graph Theory with Applications*, American Elsevier, New York, 1976; *MR***54**#117.
7. A. Bouchet, Nowhere-zero integral flows on a bidirected graph, *J. Combinatorial Theory (B)* **34** (1983), 279–282; *MR***85d**:05109.
8. U. Celmins, On cubic graphs that do not have an edge-3-colouring, Ph.D. Thesis, Waterloo, 1984.
9. A. G. Chetwynd and R. J. Wilson, Snarks and supersnarks, *The Theory and Applications of Graphs* (ed. G. Chartrand *et al.*), John Wiley, New York, 1981, pp. 215–241.
10. J. Edmonds, Minimum partition of a matroid into independent subsets, *J. Res. Nat. Bur. Standards* **69B** (1965), 67–72; *MR***32**#7442.
11. H. Fleischner, Eine gemeinsame Basis für die Theorie der Eulerschen Graphen und den Satz von Petersen, *Monatsh. Math.* **81** (1976), 267–278; *MR***55**#172.
12. J. L. Fouquet, On nowhere-zero flows in bidirected graphs, Communication at the Colloque sur la Theorie Algébrique des Graphes, Le Mans, 1984.
13. J. L. Fouquet, T. Swart and U. Celmins, Characterization and constructions of snarks, to appear.
14. J. L. Fouquet, Conjecture du 5-flot pour les graphes presque planaires, Séminaire de Mathématiques Discrètes et Applications, Grenoble, 7 November 1985.
15. D. R. Fulkerson, Blocking and anti-blocking pairs of polyhedra, *Math. Programming* **1** (1971), 168–194; *MR***45**#3222.
16. L. Goddyn, A girth requirement for the double cycle cover conjecture, *Cycles in Graphs*, *Ann. Discrete Math.* **27**, North-Holland, Amsterdam, 1985, pp. 13–26; *MR***87b**#05083.

17. H. Grötzsch, Zur Theorie der diskreten Gebilde. VII. Ein Dreifarbensatz für dreikreisfreie Netze auf der Kugel, *Wiss. Z. Martin-Luther-Univ. Halle-Wittenberg. Math.-Nat. Reihe* **8** (1958/9) 109–120; *MR*22#7113c.
18. G. Haggard, Edmonds characterization of disc embeddings, *Proc. Eighth Southeastern Conference on Combinatorics, Graph Theory, and Computing*, Utilitas Math., Winnipeg, 1977, pp. 291–302.
19. F. Jaeger, Balanced valuations and flows in multigraphs, *Proc. Amer. Math. Soc.* **55** (1976), 237–242; *MR*55#191.
20. F. Jaeger, On nowhere-zero flows in multigraphs, *Proc. Fifth British Combinatorial Conference* (ed. C. St. J. A. Nash-Williams and J. Sheehan), *Congressus Numerantium XV*, Utilitas Math., Winnipeg, 1976, pp. 373–378; *MR*52#16570.
21. F. Jaeger, Sur les flots dans les graphes et certaines valuations dans les hypergraphes d'intervalles, *Proc. Colloque Algèbre Appliquée et Combinatoire* (ed. C. Benzaken), Grenoble, 1978, pp. 189–193.
22. F. Jaeger, Flows and generalized coloring theorems in graphs, *J. Combinatorial Theory (B)* **26** (1979), 205–216.
23. F. Jaeger, Tait's theorem for graphs with crossing number at most one, *Ars Combinatoria* **9** (1980), 283–287.
24. F. Jaeger, On graphic-minimal spaces, *Combinatorics 79, Ann. Discrete Math.* **8**, North-Holland, Amsterdam, 1980, pp. 123–126.
25. F. Jaeger, On circular flows in graphs, *Proc. Colloq. Math. Soc. János Bolyai* **37** (1982), 391–402.
26. F. Jaeger, A survey of the cycle double cover conjecture, *Cycles in Graphs, Ann. Discrete Mathematics* **27**, North-Holland, Amsterdam, 1985, pp. 1–12; *MR*87b#05082.
27. F. Jaeger, On five-edge-colourings of cubic graphs and nowhere-zero flow problems (Communication at the Tenth British Combinatorial Conference, Glasgow, 1985), *Ars Combinatoria* **20B** (1985), 229–244; *MR*87f#05071.
28. F. Jaeger and T. Swart, Conjecture 1, *Combinatorics 79, Ann. Discrete Math.* **9**, North-Holland, Amsterdam, 1980, p. 305.
29. A. Khelladi, Propriétés algébriques de structures combinatoires, Thèse de Doctorat d'État, Algiers, 1985.
30. P. A. Kilpatrick, Tutte's first colour-cycle conjecture, Ph.D. Thesis, Cape Town, 1975.
31. S. Kundu, Bounds on the number of disjoint spanning trees, *J. Combinatorial Theory (B)* **17** (1974), 199–203; *MR*51#5353.
32. C. H. C. Little and R. D. Ringeisen, On the strong graph embedding conjecture, *Proc. Ninth Southeastern Conference on Combinatorics, Graph Theory, and Computing*, Utilitas Math., Winnipeg, 1978, pp. 479–487.
33. G. J. Minty, A theorem on three-coloring the edges of a trivalent graph, *J. Combinatorial Theory* **2** (1967), 164–167; *MR*34#5703.
34. M. Möller, H. G. Carstens and G. Brinkmann, Tutte's 5-flow conjecture for graphs of orientable genus ≤ 2 or nonorientable genus ≤ 4, preprint.
35. C. St. J. A. Nash-Williams, Edge-disjoint spanning trees of finite graphs, *J. London Math. Soc.* **36** (1961), 445–450; *MR*24#A3087.
36. O. Ore, *The Four-Color Problem, Pure and Applied Math.* **27**, Academic Press, New York, 1967; *MR*36#74.
37. M. Preissmann, Sur les colorations des arêtes des graphes cubiques, Thèse de Doctorat d'État, Grenoble, 1981.
38. A. Raspaud, Flots et couvertures par des cycles dans les graphes et les matroïdes, Thèse de Doctorat d'État, Grenoble, 1985.
39. P. Seymour, Sums of circuits, *Graph Theory and Related Topics* (ed. J. A. Bondy and U. S. R. Murty), Academic Press, New York, 1979, pp. 341–355.
40. P. Seymour, On multi-colourings of cubic graphs, and conjectures of Fulkerson and Tutte, *Proc. London Math. Soc. (3)* **38** (1979), 423–460.
41. P. Seymour, Nowhere-zero 6-flows, *J. Combinatorial Theory (B)* **30** (1981), 130–135.
42. P. Seymour, On Tutte's extension of the four-colour problem, *J. Combinatorial Theory (B)* **31** (1981), 82–94.

43. R. Steinberg, Grötzsch' Theorem for the Projective Plane, Research working paper **479A**, Grad. School of Business, Columbia University, 1982.
44. R. Steinberg, Tutte's 5-flow conjecture for the projective plane, *J. Graph Theory* **8** (1984), 277–289; *MR*85f:05060.
45. G. Szekeres, Polyhedral decompositions of cubic graphs, *Bull. Austral. Math. Soc.* **8** (1973), 367–387; *MR*48#3785.
46. M. Tarsi, Nowhere zero flow and circuit covering in regular matroids, *J. Combinatorial Theory (B)* **39** (1985), 346–352; *MR*87a#05049.
47. M. Tarsi, Semi duality and the cycle double cover conjecture, *J. Combinatorial Theory (B)* **41** (1986), 332–340.
48. W. T. Tutte, On the imbedding of linear graphs in surfaces, *Proc. London Math. Soc. (2)* **51** (1950), 474–483; *MR*10–616b.
49. W. T. Tutte, A contribution to the theory of chromatic polynomials, *Canad. J. Math.* **6** (1954), 80–91; *MR*15–814c.
50. W. T. Tutte, A class of Abelian groups, *Canad. J. Math.* **8** (1956), 13–28, *MR*17–708a.
51. W. T. Tutte, On the problem of decomposing a graph into n connected factors, *J. London Math. Soc.* **36** (1961), 221–230; *MR*25#3858.
52. W. T. Tutte, On the algebraic theory of graph colorings, *J. Combinatorial Theory* **1** (1966), 15–50; *MR*33#2573.
53. W. T. Tutte, The dichromatic polynomial, *Proc. Fifth British Combinatorial Conference*, *Congressus Numerantium XV*, Utilitas Math., Winnipeg, 1976, pp. 605–635; *MR*53#186.
54. P. N. Walton and D. J. A. Welsh, On the chromatic number of binary matroids, *Mathematika* **27** (1980), 1–9; *MR*81m:05060.
55. D. J. A. Welsh, *Matroid Theory*, *London Math. Soc. Monographs* **8**, Academic Press, London, 1976; *MR*55#148.
56. A. T. White, *Graphs, Groups and Surfaces*, North-Holland, Amsterdam, London, 1984; *MR*86d:05047.
57. N. H. Xuong, Sur quelques problèmes d'immersion d'un graphe dans une surface, Thèse de Doctorat d'État, Grenoble, 1977.
58. D. H. Younger, Integer flows, *J. Graph Theory* **7** (1983), 349–357; *MR*85d#05223.
59. A. Zyka, to appear.

5
Paths, Circuits and Subdivisions

CARSTEN THOMASSEN

1. Introduction

One of the most famous conjectures in graph theory is *Hadwiger's conjecture*, that every graph with chromatic number at least k contains a subgraph contractible to K_k, the complete graph on k vertices. A stronger conjecture, attributed to Hajós, asserts that every k-chromatic graph actually contains a subdivision of K_k (see Fig. 1, which shows a subdivision of K_5). However, Hajós' conjecture is false, as was shown by Catlin [**11**]; a counter-example for $k = 8$ is given in Fig. 2.

The graph in Fig. 2 is obtained by arranging five complete graphs in a cyclic order and joining each pair of consecutive graphs completely. The chromatic numbers of such graphs had previously been studied by Gallai (B. Toft, private communication). Catlin's surprising discovery was that some of these graphs do not contain subdivisions of large complete graphs. Thus the graph of Fig. 2 has chromatic number 8, but contains no subdivision of K_8; this follows since any such subdivision contains eight vertices joined, pair-by-pair, by seven internally disjoint paths, whereas in Fig. 2, for any seven vertices in the graph, some two are separated by just six vertices. Subsequently, Erdős and Fajtlowicz [**17**] obtained the

GRAPH THEORY, 3
ISBN 0–12–086203–4

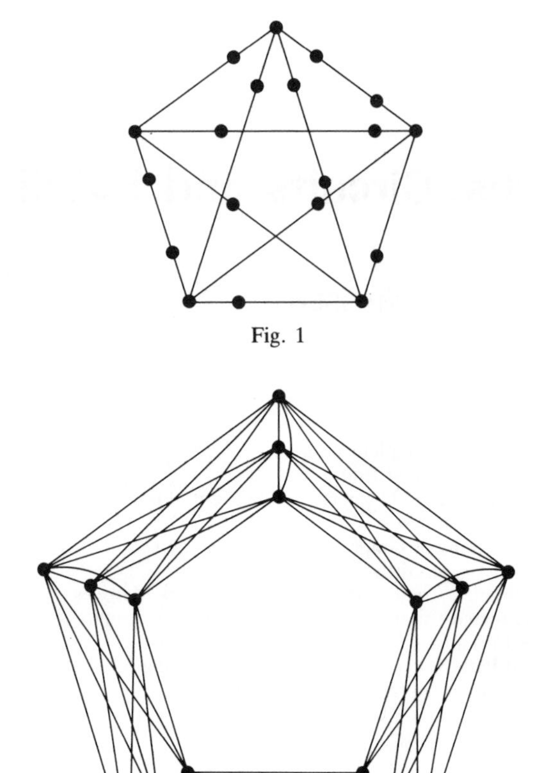

Fig. 1

Fig. 2

even more surprising result that almost all graphs are counter-examples to Hajós' conjecture.

These results add interest to the question, raised by Halin in his doctoral thesis, as to whether there exists a function $f: \mathbf{N} \to \mathbf{N}$ such that any graph with chromatic number $f(k)$ contains a subdivision of K_k. In this connection, Wagner **[70]** proved that every graph whose chromatic number is at least 2^k contains a subgraph contractible to K_k.

Wagner's elegant induction argument which requires familiarity only with the distance classes of a graph is as follows. The statement is trivial for $k = 1$ and 2, so let G be a graph with chromatic number at least 2^k, where $k \geq 3$. Let G' be a connected component of G with chromatic number at least 2^k, pick a vertex v_0 of G', and let G_i be the subgraph induced by the vertices of distance i from v_0. For some i, the subgraph G_i

has chromatic number at least 2^{k-1}, for otherwise G' can be colored in $2^k - 2$ colors (using the colors $\{1, 2, \ldots, 2^{k-1} - 1\}$ in G_i when i is odd, and the colors $\{2^{k-1}, 2^{k-1} + 1, \ldots, 2^k - 2\}$ in G_i when i is even). So, by induction, some G_i can be contracted into a graph containing K_{k-1}. Contracting $G_0 \cup G_1 \cup \ldots \cup G_{i-1}$ to a single vertex proves the statement.

Dirac **[15]** and Jung **[27]** have observed that Wagner's method can be modified to answer Halin's question in the affirmative:

Theorem 1.1. *For each natural number k, there exists a natural number $f(k)$ such that any $f(k)$-chromatic graph contains a subdivision of K_k.* ||

The purpose of this chapter is to discuss some results, methods and problems which arise from the above weakening of Hajós' conjecture. In particular, we present a variety of results on subdivisions of complete (or dense) graphs and on path systems in undirected graphs with constraints on the minimum degree, connectivity, chromatic number or girth. We also consider the possibility of extending some of these results to directed graphs, and discuss in particular the problem of finding a directed circuit of even length. Finally, we describe results of a more general nature for tournaments.

Some basic problems are stated in this chapter. In a forthcoming paper **[66]** we state a number of more specialized problems and indicate lines of possible future research in this area.

Throughout this chapter, "graph" means "simple graph", unless otherwise stated.

2. Subdivisions of Complete Graphs

In **[38]**, Mader obtained a fundamental result which extends Theorem 1.1. Before we state it, we observe that the aforementioned argument of Wagner yields the following result:

Theorem 2.1. *If G is a graph such that $\chi(G) > 2k$, where k is a natural number, then G contains two disjoint subgraphs G' and H such that*

(*i*) $\chi(G') > k$,

(*ii*) *H is connected, and*

(*iii*) *each vertex of G' is joined by an edge to H.* ||

The analogue of Theorem 2.1, with chromatic number replaced by minimum degree, is contained in the next proposition, which appears implicitly in **[38]**:

Theorem 2.2. *If G is a graph such that $\delta(G) \geq 2k$, then G contains two disjoint subgraphs G' and H such that* (*i*) $\delta(G') \geq k$,

(*ii*) *H is connected, and*

(*iii*) *each vertex of* G' *is joined by an edge to* H.

Proof. We may assume that G is connected, since otherwise we restrict our attention to a connected component of G. Consider those connected subgraphs H' of G with the property that, if we contract H' into a single vertex and replace any resulting multiple edges by single edges, then the resulting graph has at least $k(|V(G)| - |V(H')| + 1)$ edges. Since $\delta(G) \geq 2k$, G has at least $k|V(G)|$ edges, and hence each vertex of G can play the role of H. Now let H be a maximal such subgraph H', and let G' be the subgraph of $G - V(H)$ induced by those vertices which are joined to H. Any vertex v of G' must have at least k neighbors in G', for otherwise we can add v to H and obtain a contradiction to the maximality of H. So $\delta(G') \geq k$, and the proof is complete. ||

We can now state and prove Mader's theorem:

Theorem 2.3. *If* G *is a graph with* $\delta(G) \geq 2^{\binom{k}{2}}$, *then* G *contains a subdivision of* K_k.

Proof. We prove, by induction on m, the more general statement that if J is any simple graph with m edges and no isolated vertices, and if $\delta(G) \geq 2^m$, then G contains a subdivision of J.

For $m = 1$ this is trivial, so assume $m \geq 2$. Let H and G' be as in Theorem 2.2, such that $\delta(G') \geq 2^{m-1}$, and let e be any edge of J. By the induction hypothesis, G' contains a subdivision of $J - e$ (except for the isolated vertices of $J - e$, if there are any). If $J - e$ has at most one isolated vertex, then it is easy to use H to obtain a subdivision of J. If $J - e$ has two isolated vertices—that is, the ends v and w of e both have degree 1 in J—then we select any two neighboring vertices v' and w' of G and apply the induction hypothesis to obtain a subdivision of $J - \{v, w\}$ in $G - \{v', w'\}$. This completes the proof. ||

We now define $h(k)$ to be the smallest natural number such that any graph with minimum degree at least $h(k)$ contains a subdivision of K_k. It follows from Theorem 2.3 that $h(k) \leq 2^{\binom{k}{2}}$. Mader **[40]** sharpened this, and recently it has been dramatically improved by Szemerédi (private communication). In order to get a *lower* bound for $h(k)$, we observe that there is a subdivision of K_{2k} in the complete bipartite graph $K_{\frac{1}{2}k(k+1),\frac{1}{2}k(k+1)}$, but not in $K_{\frac{1}{2}k(k+1),\frac{1}{2}k(k+1)-1}$. Thus $h(2k) \geq \frac{1}{2}k(k+1)$. Szemerédi has shown that this gives essentially the order of growth of $h(k)$:

Theorem 2.4. *There exists a positive constant c such that, for all natural numbers* $k \geq 2$,

$$\tfrac{1}{8}k^2 < h(k) < ck^2 \log k. \;\|$$

The exact value of $h(k)$ is known only for small values of k. Clearly, we have $h(k) = k - 1$, for $k = 2, 3$, and Dirac [**13**] showed that this holds for $k = 4$ as well. (This will be derived from a more general result in Section 6.) Pelikán [**42**] proved that if $\delta(G) \geq 4$, then G contains a subdivision of K_5 minus one edge, and a short proof was also given by the author [**50**]. Combining this with Theorem 2.2, we easily deduce that $h(5) \leq 8$.

In [**50**] there appears the stronger result that any graph with p vertices and at least $4p - 10$ edges contains a subdivision of K_5. Dirac [**14**] has made the following conjecture:

Conjecture 2.1. *Any graph with* p *vertices and at least* $3p - 5$ *edges contains a subdivision of* K_5.

Any maximal planar graph with p vertices has $3p - 6$ edges, and thus the bound in Conjecture 2.1 is sharp. Also there are planar graphs (for example, the icosahedron graph) showing that $h(5) \geq 6$. If true, Conjecture 2.1 implies that $h(5) = 6$.

3. Paths with Prescribed Ends

We now apply the existence of $h(k)$ to path systems in graphs. We say that a graph G is **k-linked** if it has at least $2k$ vertices and, for any ordered set $\{v_1, \ldots, v_k, w_1, \ldots, w_k\}$ of $2k$ distinct vertices, G has k disjoint paths $P_1, \ldots, P_k$ such that P_i connects v_i and w_i for $i = 1, \ldots, k$. The next result on k-linked graphs was obtained independently by Jung [**28**] and Larman and Mani [**34**]; in order to obtain a short proof, we have made no attempt to minimize the bound on the connectivity:

Theorem 3.1. *If* G *is a* $2k$*-connected graph which contains a subdivision of* K_{3k}, *then* G *is* k*-linked.*

Proof. Let $v_1, \ldots, v_k, w_1, \ldots, w_k$ be any distinct vertices in G, and let K be a subdivision of K_{3k} in G. By Menger's theorem, G contains $2k$ disjoint paths $P_1, \ldots, P_k, P'_1, \ldots, P'_k$ such that P_i connects v_i with a branch vertex of K and P'_i connects w_i with a branch vertex of K, for $i = 1, \ldots, k$. We select this path system in such a way that no intermediate vertex of any P_i or P'_i is a branch vertex of K, and such that the total number of edges in the path system, but not in K, is as small as possible. Consider a branch vertex v of K which is not an end of any path P_i or P'_i. Let P''_1, P'''_1 be the paths in K from v to the ends in K of P_1 and P'_1, such that neither P''_1 nor P'''_1 has an intermediate vertex which is a branch vertex of K. The minimality property of $P_1, \ldots, P_k$, and $P'_1, \ldots, P'_k$ easily implies that P''_1 and P'''_1 do not intersect any of the paths $P_2, \ldots, P_k$, or $P'_2, \ldots, P'_k$, and hence that $P_1 \cup P'_1 \cup P''_1 \cup P'''_1$ contains a $v_1 w_1$-path intersecting none of the paths $P_2, \ldots, P'_k$. Continuing in this way we obtain the desired path system. ||

Corollary 3.2. If $v_1, \ldots, v_k, w_1, \ldots, w_k$ *are vertices in an* $h(3k)$*-connected graph* G*, then* G *has* k *paths* $P_1, \ldots, P_k$ *such that each* P_i *connects* v_i *and* w_i *and has no other vertices in common with* $\{v_1, \ldots, w_k\} \cup P_1 \cup \ldots P_{i-1} \cup P_{i+1} \cup \ldots \cup P_k$.

Proof. If v is a vertex which appears m times in the sequence $v_1, \ldots, v_k, w_1, \ldots, w_k$, then we replace v by a complete graph K_m all of whose vertices have the same neighbors in $G - v$ as v. The resulting graph G' is $h(3k)$-connected and therefore contains a subdivision of K_{3k}, by Theorem 2.3. By Theorem 3.1, G' is k-linked. But this implies the existence of $P_1, \ldots, P_k$ in G, since we may regard $v_1, \ldots, w_k$ as distinct in G'. ||

The smallest connectivity needed in order to ensure that a graph is k-linked is not known for $k \geq 3$. Jung **[28]** proved that every 6-connected graph is 2-linked, and that there are 5-connected planar graphs which are not. Subsequently, Seymour **[48]** and the author **[53]** gave a complete description of those graphs which are not 2-linked.

Corollary 3.2 implies that in an $h(3k)$-connected graph, any k vertices lie on a common circuit, and that they can occur in any prescribed order around this. A similar statement even holds for k pairwise non-adjacent edges. If we are not concerned about the cyclic ordering, we can relax the connectivity condition. Dirac **[13]**, **[14]** proved the following by a well-known simple and elegant argument:

Theorem 3.3. *Any* k *vertices in a* k*-connected graph lie on a common circuit.* ||

In **[71]** Woodall proved that any k pairwise non-adjacent edges in a $(2k + 2)$-connected graph lie on a common circuit. The author proved the same for $\lceil \frac{3}{2}k \rceil$-connected graphs (see **[25]**), and Häggkvist and Thomassen **[25]** proved the following result which was conjectured by Woodall **[71]**:

Theorem 3.4. *Any set of* k *pairwise non-adjacent edges in a* $(k + 1)$*-connected graph lie on a common cycle.* ||

If E is a set of k pairwise non-adjacent edges in a k-connected graph G, and if $G - E$ is disconnected and k is odd, then G has no circuit which includes E. Lovász **[36]** conjectured that this is the only way in which Theorem 3.4 can fail for k-connected graphs, and he verified this conjecture for $k = 3$.

We conclude this section with a remark on its algorithmic aspects. For the basic definitions and results on complexity and *NP*-completeness, the reader is referred to **[1]**. Karp **[29]** showed that it is an *NP*-complete problem to find a collection of k disjoint paths with prescribed ends, when k is not fixed. For k fixed, however, the situation is different. During the past few years, Robertson and Seymour have studied contraction properties of graphs intensively, and one important consequence of their investigations is the following result, which was announced in **[43]**; the case $k = 2$ was solved completely in **[45]**, **[48]** and **[53]**:

Theorem 3.5. *For any fixed natural number k, there exists a polynomially bounded algorithm for finding k disjoint paths with prescribed ends in a graph.* ||

Theorem 3.5 implies in particular that, if k is fixed, then there is a good algorithm for finding a subdivision of K_k in a graph. The result in **[29]** easily implies that it is *NP*-complete to determine the largest k such that an arbitrary graph contains a subdivision of K_k.

4. Subgraphs of Large Connectivity and Semi-topological Subgraphs

It is clear that any k-chromatic graph contains a subgraph with minimum degree at least $k - 1$, since we can simply take a minimal k-chromatic subgraph. It follows that Theorem 2.3 implies Theorem 1.1. The next result shows that results on connectivity can be applied to results on the minimum degree, and hence also to results on the chromatic number.

Clearly, any k-connected graph has minimum degree at least k. The converse is not true, but we have the following useful result of Mader **[39]**:

Theorem 4.1. *If G is a graph with minimum degree at least* $4k$, *then G contains a k-connected subgraph.*

Proof. For technical reasons we prove a stronger result—namely, that a k-connected subgraph can always be found in the graph G, provided that

(*i*) $|V(G)| = p \geq 2k - 1$, and

(*ii*) $|E(G)| \geq (2k - 3)(p - k + 1) + 1$.

The proof is by induction on p. If $p = 2k - 1$, then G is complete, so we assume that $p \geq 2k$. If v is a vertex with degree at most $2k - 3$, then we can apply the induction hypothesis to $G - v$ and so can assume that $\delta(G) \geq 2k - 2$. If G is k-connected, then there is nothing to prove. So we can assume that G has two subgraphs G_1 and G_2 such that $G = G_1 \cup G_2$, $|V(G_1) \cap V(G_2)| = k - 1$, and $|V(G_i)| < |V(G)|$ for $i = 1, 2$. If w is a vertex in $V(G_i)\backslash V(G_{3-i})$, then w has degree at least $2k - 2$, and hence

$|V(G_i)| \geq 2k - 1$. We can now complete the proof by applying the induction hypothesis to either G_1 or G_2, since otherwise,

$$|E(G_i)| \leq (2k - 3)(|V(G_i)| - k + 1)$$

for $i = 1, 2$, and hence

$$|E(G)| \leq |E(G_1)| + |E(G_2)| \leq (2k - 3)(|V(G_1)| + |V(G_2)| - 2k + 2)$$
$$= (2k - 3)(p - k + 1),$$

which is a contradiction. ||

Using Theorems 3.1 and 4.1, we can extend Theorem 2.3 to what Bollobás **[7]**, **[8]** calls *semi-topological subgraphs*. These are subdivisions of graphs where certain prescribed edges are not subdivided. Suppose that we wish to find a subdivision of K_k in a graph of large connectivity, minimum degree or chromatic number (depending only on k), and suppose that some prescribed edges in K_k are not subdivided at all. Then these prescribed edges cannot form a circuit, since it follows from a result of Erdős **[16]** that, for each natural number m, there exists an m-chromatic graph with girth greater than k. On the other hand, Bollobás **[7]** has proved the following extension of Theorem 2.3, for which we give a short proof.

Theorem 4.2. *Let F be a spanning forest in K_k. If G is any graph of minimum degree at least $4h(3\binom{k}{2})$, then G contains a subdivision of K_k for which the edges of F are not subdivided.*

Proof. By Theorem 4.1, G contains an $h(3\binom{k}{2})$-connected subgraph H. In particular, $\delta(H) \geq h(3\binom{k}{2}) > k$, and hence H contains a copy of F as a subgraph. Now Corollary 3.2 (with $\binom{k}{2}$ instead of k) shows that F can be extended to a subdivision of K_k, and the result follows. ||

As mentioned earlier, Dirac proved that if $\delta(G) \geq 3$, then G contains a subdivision of K_4. A stronger result was proved by Toft and the author **[67]**:

Theorem 4.3. *If G is a graph with minimum degree at least 3, then G contains a subdivision of K_4 with the property that a Hamiltonian path of K_4 is not subdivided.* ||

The proof of Theorem 4.3 is based on a decomposition result which we discuss in Section 6. In **[51]**, it was shown that any graph with n vertices and at least $2n - 2$ edges contains a circuit C, and a vertex not on C which is joined to at least three vertices of C. In particular, $\delta(G) \geq 4$ implies such a configuration, and hence the term $4h(3\binom{k}{2})$ in Theorem 4.2 can be replaced by 4 when $k = 4$. An infinite class of graphs based on $K_{3,3}$ (see **[51]**) show that $\delta(G) \geq 3$ is not sufficient. However, the following result appeared in **[67]**:

Theorem 4.4. *Every 4-chromatic graph contains a subdivision of K_4 with the property that any prescribed forest in K_4 is not subdivided.* ||

Theorem 4.3 shows that every graph with minimum degree 3 contains a circuit with two crossing chords. For 3-connected graphs we can do better. The following result was conjectured by Kelmans, and proved independently by Kelmans **[30]** and the author **[60]**:

Theorem 4.5. *Every 3-connected non-planar graph, other than K_5, contains a circuit with three pairwise crossing chords.* ||

Since the configuration in Theorem 4.5 is a special subdivision of $K_{3,3}$, Theorem 4.5 gives a new planarity criterion for 3-connected graphs. As shown in **[60]**, it implies a refinement of Kuratowski's theorem for all graphs.

5. Subdivisions modulo *k*

The remark preceding Theorem 4.2 shows that a large (fixed) chromatic number does not imply the presence of a subdivision of a subgraph $K_m (m \geq 3)$ for which each edge is subdivided a prescribed number of times. But perhaps the following weaker statement holds:

Conjecture 5.1. *If G is a graph with minimum degree at least 10^{10}, then G contains a subdivision of K_4 with the property that all six edges of K_4 are subdivided the same number of times.*

It is tempting to generalize Conjecture 5.1 to K_m instead of K_4. The following result was conjectured by Bollobás **[9]** and proved by the author **[59]**:

Theorem 5.1. *For any natural numbers d and m, there exists a natural number $q(d, m)$ such that any graph G with minimum degree $q(d, m)$ contains a subdivision of some graph H with minimum degree $\delta(H) \geq d$, such that each edge of H is subdivided precisely m times.* ||

The proof of Theorem 5.1 is based on the following domination result in bipartite graphs **[59]**:

Theorem 5.2. *Let G be a bipartite graph with partite sets A and B, and suppose that each vertex of B has degree at least $k^4 + k^2$. Then there are subsets A' and B' of A and B, respectively, such that*

$$|A'| \leq |A|/k, \ |B'| \leq |B|/k,$$

and each vertex of $B \backslash B'$ is adjacent to at least k vertices of A'. ||

Roughly speaking, Theorem 5.2 says that A contains a small subset A' such that almost all vertices of B have many neighbors in A'.

It is easy to see that every graph with minimum degree at least 3 contains a circuit of even length. One way of seeing this is to use a subdivision of K_4, which is present by Dirac's theorem (which is a special case of Theorem 4.3). In such a subdivision there are three internally disjoint paths with the same ends. Two of these have the same parity, and hence we get an even circuit.

Burr and Erdős (see **[9]**) raised the following more general question. If d and k are natural numbers, does there exist a natural number $p(d, k)$ such that every graph with minimum degree at least $p(d, k)$ contains a circuit of length $2d$ (modulo k)? (The reason for writing $2d$ instead of d is that the complete bipartite graphs can have large degrees but no odd circuits.) Burr and Erdős used Theorem 2.3 to demonstrate the existence of $p(d, k)$ for special values of d and k. For $d = 0$, the existence of $p(d, k)$ follows from Theorem 2.3, combined with Theorem 5.1. Bollobás **[6]** settled the question completely by showing that a graph with large minimum degree contains $k + 2$ paths $P_1, \ldots, P_{k+2}$ such that $P_1, \ldots, P_{k+1}$ start at the same vertex v_0 and have length d, P_{k+2} contains the end of P_i (other than v_0) for $i = 1, \ldots, k$, and $|V(P_i) \cap V(P_j)| = 1$ for $1 \leq i < j \leq k + 2$. We shall refer to $P_1 \cup \ldots \cup P_{k+2}$ as a **(*k*, *d*)-fan**; a (6, 7)-fan appears in Fig. 3.

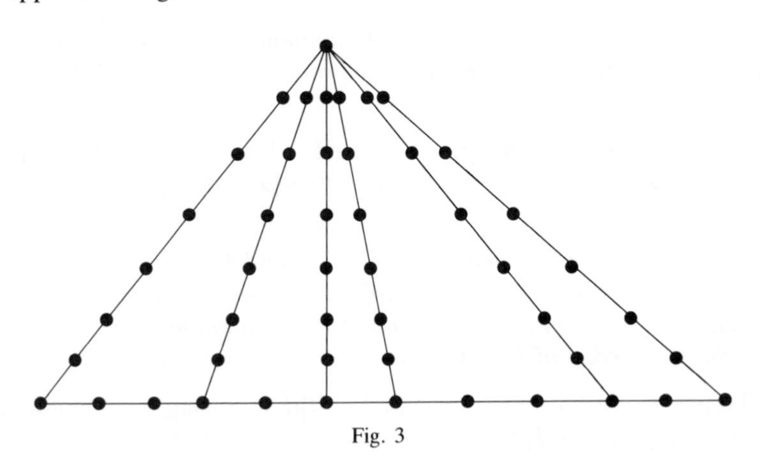

Fig. 3

It is easy to see that, for each set of $k + 1$ natural numbers, there are two numbers whose difference is divisible by k. Considering the vertices of P_{k+2} above as natural numbers we see that some subpath of P_{k+2}, together with two of the paths $P_1, \ldots, P_{k+1}$, forms a circuit of length $2d$ (mod k). In **[6]**, Bollobás proved directly that a graph with large minimum degree contains a (k, d)-fan, and in **[7]** this result was also obtained from Theorem 4.2. Here we give a short proof based on Theorems 3.3 and 4.1:

Theorem 5.3. *If G is a graph with minimum degree at least $4d(k + 1)$, then G contains a (k, d)-fan.*

Proof. By Theorem 4.1, G contains a $d(k + 1)$-connected subgraph H. We select a vertex in H and grow $k + 1$ paths $P_1, \ldots, P_{k+1}$ having only v_0 in common, pair by pair. The minimum degree of H enables us to continue until $P_1, \ldots, P_{k+1}$ all have length d. If v_i is the end of P_i other than v_0, then $F = (P_1 - v_1) \cup \ldots \cup (P_{k+1} - v_{k+1})$ has $(k + 1)(d - 1) + 1$ vertices, and so $H - V(F)$ is k-connected. By Theorem 3.3, $H - V(F)$ has a path P_{k+2} containing $v_1, \ldots, v_{k+1}$, and thus $P_1, \ldots, P_{k+2}$ form a (k, d)-fan. ||

Corollary 5.4. *If G is a graph with minimum degree $4k(k + 1)$, then G contains circuits of all even lengths modulo k.*

Proof. The set $\{2, 4, \ldots, 2k\}$ is a complete system of residues modulo $2k$. ||

The next result **[56]** generalizes several of the previous results, except for the actual values of the functions:

Theorem 5.5. *For each pair of natural numbers m and k, there exists a natural number $r(m, k)$ with the property that every graph with minimum degree at least $r(m, k)$ contains a subdivision F of $K_{m,m}$ such that each edge of $K_{m,m}$ corresponds to a path of length 1 (modulo k) in F.* ||

Theorem 5.5 implies the following extension of Theorem 2.3:

Theorem 5.6. *If G is any graph with minimum degree $r(m(m + 1), k)$, and if we assign an even natural number $\alpha(e)$ to each edge of K_m, then G contains a subdivision of K_m such that each edge e in K_m corresponds to a path of length $\alpha(e)$ (modulo k) in G.*

Proof. By Theorem 5.5, G contains a subdivision F of $K_{m(m+1), m(m+1)}$ (with partite sets A and B, say) such that each edge in this complete bipartite graph corresponds to a path of length 1 (modulo k). Select m vertices in A; these correspond to the vertices of K_m. Any two of these vertices can be joined by a path in F which has any prescribed even length (modulo k) and which uses at most $2k$ vertices of B. This proves Theorem 5.6. ||

We also get a counterpart to Theorem 3.1 and Corollary 3.2:

Theorem 5.7. *Let k be an odd natural number, G be an $r(m + 2k\binom{m}{2}, k)$-connected graph containing vertices $v_1, \ldots, v_m, w_1, \ldots, w_m$, and α_i be a natural number for $i = 1, \ldots, m$. Then G has m pairwise disjoint paths $P_1, \ldots, P_m$ such that P_i connects v_i and w_i and has length α_i (modulo k), for $i = 1, \ldots, m$.*

Proof. By Theorem 5.5, G has a subdivision F of $K_{m+2k\binom{m}{2},m+2k\binom{m}{2}}$ such that all edges of this bipartite graph correspond to paths of lengths 1 (modulo k) in G. As in the proof of Theorem 3.1, we consider $2k$ disjoint paths from $\{v_1, \ldots, v_k, w_1, \ldots, w_k\}$ to the branch vertices of F, using as few edges of F as possible. By using the $4k\binom{m}{2}$ branch vertices of F which are not used by this path system we can extend the path system to the desired paths $P_1, \ldots, P_m$. ||

The parity condition on $\alpha(e)$ in Theorem 5.6 and on k in Theorem 5.7 cannot be omitted, as shown by the complete bipartite graphs. But for graphs with large chromatic number we can avoid that condition, as shown in Theorem 5.9 below. In **[56]**, the proof of that result was based on the following counterpart to Theorem 4.1:

Theorem 5.8. *For each natural number k, there exists a natural number $t(k)$ such that each $t(k)$-chromatic graph contains a k-connected graph with chromatic number at least k.* ||

The proof of Theorem 5.8 given in **[56]** is in error, and is set correct in **[4]**. Here we prove Theorem 5.9 (below) without using Theorem 5.8. Instead, we use an argument similar to Wagner's argument in connection with Theorem 1.1.

Theorem 5.9. *Let m, k and r be natural numbers with $m \geq 2r$. Then there exists a natural number $s(m, k, r)$ such that each graph G with chromatic number at least $s(m, k, r)$ contains a subdivision of the graph J obtained from $K_{m,m}$ by adding r independent edges in the same partite set, such that each edge of J corresponds to a path of length* 1 *(modulo k) in G.*

Proof. The proof is by induction on r. For $r = 0$, the statement follows from Theorem 5.6. If $r \geq 1$, we claim that $s(m, k, r) \leq 2s(m + k, k, r - 1)$. For, if $\chi(G) \geq 2s(m + k, k, r - 1)$, then choose a component G' with the same chromatic number as G, and select a vertex v_0 of G'. It follows that some distance class from v_0 induces a graph H whose chromatic number is at least $s(m + k, k, r - 1)$. By the induction hypothesis, H contains a subdivision F of $K_{m+k,m+k}$ with $r - 1$ additional edges, as described in the statement of the theorem. We modify F to give the desired subdivision of $K_{m,m}$ with r additional edges. Let A be the partite set of $K_{m+k,m+k}$ to which we have added the $r - 1$ edges, and let v, w be two vertices in A which are not incident with any of these $r - 1$ edges. Let z be a vertex in F which is a neighbor of v. Now we consider two shortest paths from v_0 to w and z, respectively. Since these contain a path of even length from w to z, we get a path P of odd length from v to

w. If this length is 1 (modulo k), we have finished. Otherwise, we use F to extend P to a path of length 1 (modulo k) connecting two vertices of A. (We leave the details of this argument to the reader.) $\|$

Theorem 5.9 implies the following result:

Theorem 5.10. *If G is a graph with chromatic number at least $s(m\binom{m}{2}, k, \binom{m}{2})$, and we assign a natural number $\alpha(e)$ to each edge e of K_m, then G has a subdivision of K_m with the property that each edge e in K_m corresponds to a path of length $\alpha(e)$ (modulo k) in G.* $\|$

Theorem 5.10 shows in particular that a graph of large chromatic number contains circuits of all lengths modulo k. But here we can do better:

Theorem 5.11 *If G is a 2-connected non-bipartite graph, and if $\delta(G) \geq h(2k, k)$, then G has circuits of all lengths modulo k.* $\|$

The proof of Theorem 5.11 in based on Theorem 5.5, and is given in **[56]**.

One of the simplest unsolved problems in connection with Theorem 5.10 is the following conjecture of Toft **[68]**:

Conjecture 5.2. *Every 4-chromatic graph contains a subdivision of K_4 with the property that each edge of K_4 corresponds to a path of odd length in the subdivision.*

6. Decompositions of Graphs

Theorems 2.1 and 2.2 are examples of decomposition results which are useful for investigations on subdivisions. In this section we focus on that type of decomposition. We begin with a remarkable observation of P. Erdős:

Theorem 6.1. *Every graph G (which may contain multiple edges) contains a spanning bipartite graph H such that, for each vertex v of G,*

$$d_H(v) \geq \tfrac{1}{2}d_G(v).$$

Proof. Let H be a spanning bipartite subgraph of G with as many edges as possible. If some vertex v does not satisfy the inequality of Theorem 6.1, we delete it from the partite set of H to which it belongs and add it to the other partite set. This yields a bipartite graph with more edges, and gives the required contradiction. $\|$

Theorem 6.1 says, in particular, that $|E(H)| \geq \frac{1}{2}|E(G)|$, and $\delta(H) \geq$

$\frac{1}{2}\delta(G)$. This shows that our earlier results on subdivisions in a graph of large minimum degree cannot be improved (except, perhaps, for a multiplicative factor $\frac{1}{2}$ in the functions), even if we consider only bipartite graphs. Let us also mention some positive consequences of Theorem 6.1:

Theorem 6.2. *Every digraph D with p vertices and at least $4p$ arcs contains an antidirected circuit—that is, a circuit where the arcs alternate in direction.*

Proof. Let G be the underlying graph of D. By Theorem 6.1, G has a bipartite subgraph H with partite sets A and B, say, with at least $2p$ edges. Without loss of generality, we can assume that at least p arcs of D are directed from A to B, and the subgraph of H consisting of these edges contains a circuit which in D is antidirected. ||

Theorem 6.2 implies, in particular, that every digraph with minimum total degree at least 8 has an antidirected cycle. This can be improved, as shown in [**19**].

Theorem 6.1 also implies the following result of Graham [**20**]:

Theorem 6.3. *If T is an oriented tree with no directed path of length 2, then there exists a natural number c_T with the property that every digraph with minimum (total) degree at least c_T contains a copy of T.*

Proof. We show that $c_T \leq 8m$, where m is the maximum degree of T. For if the minimum degree of D is at least $8m$, then D has at least $4m|V(D)|$ arcs. As in the proof of Theorem 6.2, Theorem 6.1 implies that there exists a subdigraph D' in D with at least $m|V(D)|$ arcs, all of which are directed from one of the partite sets to the other. If D'' is a minimal subdigraph of D' with at least $m|V(D'')|$ arcs, then D'' has minimum degree at least m. Since T is a tree, it is bipartite with partite sets A and B (say), and since T has no directed path of length 2, we can assume that all arcs of T are directed from A to B. Also, D'' is bipartite with partite sets A' and B' (say), such that all arcs of D'' are directed from A' to B'. It is now easy to embed T in D'' such that $A \subseteq A'$ and $B \subseteq B'$. ||

The proof of Theorem 6.3 shows that any digraph D with at least $4|V(T)|\,|V(D)|$ arcs contains a copy of T.

We now discuss decomposition problems where we are primarily interested in conditions on the two parts, rather than the edges between the two parts as in Theorem 6.1. Tutte [**69**] showed that every 3-connected graph has a chordless circuit whose deletion leaves a connected graph. The following result in [**67**] is slightly stronger:

Theorem 6.4. *If G is a connected graph and $\delta(G) \geq 3$, then G has a chordless circuit C such that $G - V(C)$ is connected.*

Proof. Among all circuits of G we select one (C, say), such that

(i) the largest component H of $G - V(C)$ has as many vertices as possible, and

(ii) subject to (i), $|V(C)|$ is as small as possible.

Clearly (ii) implies that C is chordless. We shall show that $G - V(C)$ is connected. Suppose therefore, to obtain a contradiction, that $G - V(C)$ has a component $H' \neq H$. Then H' cannot have a circuit. For if C' were a circuit in H', then $G - V(C')$ would have a component containing H and C, contradicting (i). So H' is a tree. Let P be a path in H' connecting two end-vertices v_1, v_2 of H' (possibly, $v_1 = v_2 = P = H'$), and let H'' be the union of C, P and all edges from v_1 or v_2 to C. Since $\delta(G) \geq 3$, it follows easily that $H'' - w$ has a circuit C_w for each vertex w of H''. Since G is connected, C has a vertex w which is joined to H, and now C_w contradicts (i). ||

Theorem 6.4 easily implies the result that any graph G with minimum degree at least 3 contains a subdivision of K_4, such that some two adjacent edges of K_4 are not subdivided. For we can assume that G is connected, and we let C be as in Theorem 6.4. We select three consecutive vertices v_1, v_2, v_3 on C, and since C is chordless we can find three vertices w_1, w_2, w_3 in $G - V(C)$ which are neighbors of v_1, v_2, v_3, respectively. Since $G - V(C)$ is connected, it contains a path P_1 from w_1 to w_2, and a path P_2 from w_3 to P_1. Then $C \cup P_1 \cup P_2 \cup \{v_1w_1, v_2w_2, v_3w_3\}$ is a subdivision of K_4.

A refinement of Theorem 6.4 yields the stronger result given in Theorem 4.3.

E. Györi conjectured that a graph of large connectivity can be decomposed into graphs of large connectivity. This was proved in [**55**]. The proof is based on the analogous result for minimum degree which was also proved in [**55**] and subsequently improved by Häggkvist, Alon (private communication) and Hajnal [**23**], as shown in the next result:

Theorem 6.5. *Let G be a graph with minimum degree at least $2s + t - 1$, where s and t are natural numbers. Then $V(G)$ can be decomposed into non-empty sets A and B such that*

$$\delta(G(A)) \geq s \text{ and } \delta(G(B)) \geq t.$$

Proof. Among all non-empty subsets of $V(G)$ we select one, A (say), such that

(i) $|E(G(A))| \geq (s - 1)|A|$,

(ii) $|A|$ is minimum subject to (i), and

(*iii*) $|E(G(A))|$ is maximum subject to (*i*) and (*ii*).

Since $V(G)\backslash\{v\}$ satisfies (*i*), for each vertex v, A exists and is a proper subset of $V(G)$. Put $B = V(G)\backslash A$. Then $\delta(G(A)) \geq s$, for if v is in A and has degree at most $s - 1$ in $G(A)$, then $A\backslash\{v\}$ satisfies (*i*), contradicting (*ii*). If all vertices in $G(A)$ have degree at least $2s - 1$, then $A\backslash\{v\}$ satisfies (*i*) for any vertex v in A, and so A has a vertex w of degree at most $2s - 2$ in $G(A)$. If z is a vertex in B with degree at most $t - 1$ in $G(B)$, then z is joined to at least $2s$ vertices in A, and hence $(A\backslash\{w\}) \cup \{z\}$ contradicts (*iii*). This proves that $\delta(G(B)) \geq t$. ||

It is possible that Theorem 6.5 can be extended as follows:

Conjecture 6.1. *If s and t are natural numbers, and if G is a graph with minimum degree $s + t + 1$, then $V(G)$ can be partitioned into non-empty sets A and B such that*

$$\delta(G(A)) \geq s \text{ and } \delta(G(B)) \geq t.$$

The complete graph K_{s+t+1} shows that $s + t + 1$ cannot be replaced by $s + t$ in Conjecture 6.1. Using a refinement of the argument in Theorem 6.1, Lovász **[35]** proved a counterpart to Conjecture 6.1: if the maximum degree of G is at most $s + t + 1$, then G can be decomposed (as in Conjecture 6.1) into graphs with maximum degree at most s and t, respectively.

We can now prove the aforementioned conjecture of Györi.

Theorem 6.6. *If G is a $12k$-connected graph, then $V(G)$ has a partition into non-empty sets A and B, such that $G(A)$ and $G(B)$ are both k-connected.*

Proof. By Theorem 6.5, $V(G)$ has a decomposition into sets A' and B', such that $\delta(G(A')) \geq 4k$ and $\delta(G(B')) \geq 4k$. By Theorem 4.1, there are subsets $A'' \subseteq A'$ and $B'' \subseteq B'$, such that $G(A'')$ and $G(B'')$ are both k-connected. Now let A, B be sets such that

(*i*) $A'' \subseteq A$, $B'' \subseteq B$ and $A \cap B = \emptyset$,

(*ii*) $G(A)$ and $G(B)$ are both k-connected, and

(*iii*) $A \cup B$ is maximal subject to (*i*) and (*ii*).

We complete the proof by showing that $A \cup B = V(G)$. For suppose that $D = V(G)\backslash(A \cup B)$ is non-empty. By the maximality of $A \cup B$, $G(A \cup D)$ is not k-connected, and has therefore a separating set S with at most $k - 1$ vertices. Since $G(A)$ is k-connected, $G(A) - S$ is connected and hence $G(A \cup D) - S$ has a component with vertex-set $D' \subseteq D$. Again, the maximality of $A \cup B$ implies that $G(B \cup D')$ has a separating set T with at most $k - 1$ vertices, and $G(B \cup D') - T$ has a component with vertex-set $D'' \subseteq D'$. Now $G(D'')$ is a component of $G - (S \cup T)$, and so $S \cup T$ is a separating set of G with at most $2k - 2$ vertices. This contradicts the fact that G is $12k$-connected. ||

Theorem 6.6 is probably not best possible. However, the following conjecture is, if true, best possible:

Conjecture 6.2. *Conjecture* 6.1 *remains valid if "minimum degree" is replaced by "connectivity".*

Conjecture 6.2 is true for $s = 2$, as is shown by the following result of the author **[54]**; it was conjectured by Lovász **[37]**:

Theorem 6.7. *If G is a $(k + 3)$-connected graph, then G contains a circuit C such that $G - V(C)$ is k-connected.* ||

Clearly Theorem 6.7 is true if G has a triangle. For triangle-free graphs the proof is based on the following contraction result in **[54]**:

Theorem 6.8. *Every triangle-free graph contains an edge whose contraction preserves the connectivity.* ||

It would be useful to have a result stating that the vertex-set of a graph G with large connectivity or minimum degree has a decomposition into sets A and B, such that each of the graphs $G(A)$, $G(B)$ and the bipartite graph consisting of all edges between A and B in some sense has large connectivity or edge density. In **[56]** the following conjecture was made:

Conjecture 6.3. *For each natural number k, there exists a natural number $\alpha(k)$ such that, for any $\alpha(k)$-connected graph G and any subset $S \subseteq V(G)$ with cardinality at most k, $V(G)$ has a partition into sets A and B, such that $S \subseteq A$, $G(A)$ and $G(B)$ are both k-connected, and each vertex of A is joined to at least k vertices of B.*

In **[56]** it was pointed out that Conjecture 6.3 implies the following conjecture of Lovász **[37]**, which is closely related to Theorem 6.7:

Conjecture 6.4. *For each natural number k, there exists a natural number $\beta(k)$ such that, for any two vertices v, w in any $\beta(k)$-connected graph G, there exists a $v - w$ path P such that $G - V(P)$ is k-connected.*

Indeed, it is easy to see that $\beta(k) \leq \alpha(k)$, and for this we need only Conjecture 6.3 in the special case $S = \{v, w\}$.

If Conjecture 6.4 holds in the stronger version where P is an induced path avoiding a prescribed set of at most k vertices, then that conjecture, together with Theorem 4.1, would give a new proof of Theorem 2.3

(except for the numerical part). In Section 12 we point out how this idea can be applied to subdivisions in tournaments.

7. Graphs of Large Girth

Several authors have considered the problem of constructing graphs with prescribed order and minimum degree or connectivity, and with as large a girth as possible (see [**8**, Chapter 3]). Such graphs have applications to communication networks, but they are also interesting in their own right.

In [**57**] it was demonstrated that graphs of large girth and minimum degree at least 3 share many properties with graphs of large minimum degree. A basic result is:

Theorem 7.1. *If G is a graph of minimum degree at least 3 and girth at least $2k - 3$ (where k is a natural number ≥ 3), then G can be contracted into a multigraph H with minimum degree at least k, such that no two vertices of H are joined by more than two edges.*

Proof. If $k = 3$, we put $H = G$, so assume that $k \geq 4$. We can also assume without loss of generality that G is connected. Now consider a partition of the vertex-set of G into sets $A_1, \ldots, A_m$ such that $|A_i| \geq k - 2$ and $G(A_i)$ is connected for each $i = 1, \ldots, m$. Clearly, such a partition exists with $m = 1$. Among all such partitions, choose one such that m is maximum.

Let T_i be a spanning tree of $G(A_i)$, for $i = 1, \ldots, m$. We first show that $G(A_i) = T_i$, for each $i = 1, \ldots, m$. For, if $G(A_i)$ has an edge e not in T_i, then $T_i \cup \{e\}$ has a unique circuit C_i which has length at least $2k - 3$, and hence T_i has an edge e' such that $T_i - e'$ consists of two trees with vertex-sets A_i', A_i'', respectively, each of cardinality at least $k - 2$. But then the partition $A_1, \ldots, A_{i-1}, A_i', A_i'', A_{i+1}, \ldots, A_m$ contradicts the maximality of m.

We next show that no two trees T_i and T_j ($1 \leq i < j \leq m$) are joined by more than two edges. For, if e_1, e_2, e_3 are three edges joining T_i and T_j, then $T_i \cup T_j \cup \{e_1, e_2, e_3\}$ contains three internally disjoint paths P_1, P_2, P_3. Since G has girth at least $2k - 3$, we can assume that P_1 and P_2 both have length at least $k - 1$. Now let A_i' and A_i'' be $k - 2$ consecutive internal vertices of P_1 and P_2, respectively. Then each of the three sets A_i', A_i'' and $(A_i \cup A_j)\backslash(A_i' \cup A_i'')$ induces a connected subgraph of G of order at least $k - 2$. By considering these sets, instead of A_i and A_j, we obtain a contradiction to the maximality of m.

The graph H obtained by contracting each T_i ($i = 1, \ldots, m$) to a vertex satisfies the conclusion of the theorem. $\|$

Theorem 2.3 implies in particular that a graph with large minimum degree contains a subgraph which can be contracted into K_k. Kostochka [33] showed that having minimum degree $ck \log k$ implies the existence of such a subgraph (where c is a positive constant), and that this is best possible except for the value of c. Thus Kostochka's result is the 'contraction analogue' of Theorem 2.4.

Combining Kostochka's result with Theorem 7.1 we get the following:

Corollary 7.2. *If G has minimum degree at least* 3 *and girth at least* $4ck \log k$, *then G has a subgraph which can be contracted into* K_k. ||

Corollary 7.2 shows, for example, that a graph of large girth, and minimum degree at least 3, contains many disjoint circuits. But for disjoint circuits we can do considerably better. It turns out, surprisingly, that only a modest girth condition is needed. It is easy to prove that a graph with minimum degree at least $3k - 1$ contains k disjoint circuits, for we can simply delete a shortest circuit and use induction. The same argument shows that a graph with girth at least 4 and minimum degree at least $2k$ contains k disjoint circuits. These results are best possible, as shown by the complete graphs and the complete bipartite graphs, respectively. Continuing like this, we can show that any graph with girth at least 5 and minimum degree at least k contains k disjoint circuits, but this result can be improved drastically:

Theorem 7.3. *Let k be any natural number. If G is a graph of order p with minimum degree at least* 4, *girth at least* 5, *and such that* $p/(\log p)^4 > 2^{13} (k - 1)^2$, *then G contains a collection of k pairwise disjoint circuits of the same length.* ||

The proof of Theorem 7.3, in [57], shows that the circuits in that theorem can be obtained by successively deleting a shortest circuit until we get a forest. Then some k of these disjoint circuits must have the same length. Examples in [57] show that Theorem 7.3 is not valid for graphs with minimum degree 3. For such graphs we cannot even be sure that there are five disjoint circuits. For graphs of minimum degree 3, we get the following result:

Theorem 7.4. *For each natural number k, there exists a natural number* n_k *such that each graph with minimum degree at least* 3, *girth at least* 7, *and order at least* n_k, *contains k disjoint circuits of the same length.* ||

Examples in [57] show that the number 7 cannot be replaced by 6 in Theorem 7.4. Theorem 7.1 and Corollary 7.2 show that results on contraction properties of graphs with large minimum degree can be applied to graphs with large girth and minimum degree. This also goes the other way round. Häggkvist [24] was the first to show that a graph with large minimum degree contains many disjoint circuits of the same length. Using Theorem 7.3, the following strengthening of Häggkvist's result was proved in [57]:

Theorem 7.5. *For each natural number k, there exists a natural number m_k such that any graph G with order $p \geq m_k$ and minimum degree at least $3k + 1$ contains a collection of k pairwise disjoint circuits of the same length.* ||

The above-mentioned connection becomes more striking if the following conjecture holds:

Conjecture 7.1. *For each pair of natural numbers r and g, there exists a natural number $\gamma(r, g)$ such that every graph with minimum degree at least $\gamma(r, g)$ contains a subgraph with minimum degree r and girth at least g.*

The remarks preceding Theorem 7.3 were improved considerably by Corradi and Hajnal [12]:

Theorem 7.6. *Every graph with at least $3k$ vertices and minimum degree at least $2k$ contains k pairwise disjoint circuits.* ||

Theorems 7.5 and 7.6 suggest the following conjecture made in [57] and, for $k = 2$, in [24]:

Conjecture 7.2. *For each natural number k, there exists a natural number n_k such that every graph with at least n_k vertices and minimum degree at least $2k$ contains k disjoint circuits of the same length.*

Returning to graphs of large girth, we find that Theorem 4.1 also has an analogue for such graphs when an appropriate connectivity concept is considered. We say that a 2-connected graph is **cyclically k-vertex-connected** ($k \geq 2$) if it is not the union of two subgraphs G_1 and G_2, such that each of G_1 and G_2 has a circuit and $|V(G_1) \cap V(G_2)| \leq k - 1$. The next two results show that this definition is reasonable. Theorem 7.7 below is analogous to Menger's theorem, and Theorem 7.8 is a counterpart to Theorem 3.4:

Theorem 7.7. *If G has minimum degree at least 3 and is cyclically k-vertex-connected, where $k \geq 3$, and if A, B are disjoint sets of vertices of G each of cardinality at least $2k - 3$, then G contains a collection of k disjoint paths from A to B.* ||

Theorem 7.8. *If A is a set of k independent edges in a cyclically 2^{k+1}-vertex-connected graph G with minimum degree at least 3, then G has a circuit through A.* ||

The following conjecture was made in [**26**], and proved for $k \leq 4$ in [**2**]:

Conjecture 7.3. *If G is a trivalent cyclically $(k + 1)$-edge-connected graph and if A is a set of k independent edges in G, then G has a circuit through A.*

For trivalent graphs, the above definition of cyclic k-vertex-connectivity is closely related to the cyclic edge-connectivity. Indeed, a cyclically k-vertex-connected graph has girth at least k, and is cyclically k-edge-connected as well. Conversely, it is not difficult to show that a trivalent cyclically k-edge-connected graph of large order is also cyclically k-vertex-connected.

The following analogue of Theorem 4.1 was proved in [**57**]:

Theorem 7.9. *Let k be a natural number with $k \geq 2$. If G is a graph of minimum degree at least 3 and girth at least $4k - 6$, then G contains an induced subgraph H which can be described as a graph obtained from a cyclically k-vertex-connected graph H^* with minimum degree at least 3 by inserting at most $2k - 3$ vertices of degree 2 on distinct edges.* $\|$

The usefulness of Theorem 7.9 can be demonstrated by applying it in the proof of the following counterpart to Corollary 5.4:

Theorem 7.10. *If G is a graph with minimum degree at least 3 and girth at least $2(k^2 + 1)(3 \cdot 2^{k^2+1} + (k^2 + 1)^2 - 1)$, then G contains circuits of all even lengths modulo k.* $\|$

Theorems 7.4 and 7.10 show, in particular, that if we forbid the presence of small circuits in a graph with minimum degree at least 3, then we force a richness of other circuits. Another result along these lines, but proved by completely different methods, is the result in [**22**] that a graph with minimum degree at least 3 and large girth contains circuits of many different lengths. An analogous result for graphs of large minimum degree was obtained by Gyárfás, Komlós and Szemerédi [**21**].

8. Even Directed Circuits and Sign Non-singular Matrices

In the previous sections we have seen that conditions on the minimum degree, connectivity or girth imply the presence of a number of configurations (path systems, circuits or subdivisions of dense graphs) in an undirected graph. We now discuss the possibility of extending some of these results to digraphs. Before doing so, however, we demonstrate that the detection of a very simple configuration—namely, an even directed circuit (that is, a directed circuit of even length) in a digraph—is of considerable interest. It is of importance in recognizing the so-called *sign non-singular matrices*, as we discuss in this section, and in the characterization of the 3-color-critical hypergraphs with the smallest possible number of edges (discussed in the next section).

We say that two real $m \times n$ matrices $\mathbf{A} = (a_{ij})$ and $\mathbf{A}' = (a'_{ij})$ have the *same sign pattern* if, for all i and j, $a_{ij} = a'_{ij} = 0$ or $a_{ij}a'_{ij} > 0$. In certain problems in qualitative linear algebra arising from mathematical economics (see [**44**]), one is interested in a linear system of equations of the form $\mathbf{Ax} = \mathbf{b}$ (where $\mathbf{b} \in \mathbf{R}^m$ and $\mathbf{x} \in \mathbf{R}^n$), where the sign pattern of the solutions $\mathbf{x}$ depend only on $\mathbf{A}$ and $\mathbf{b}$. More precisely, if $\mathbf{A}'$ and $\mathbf{b}'$ have the same sign pattern as $\mathbf{A}$ and $\mathbf{b}$, respectively, then the solutions of $\mathbf{A}'\mathbf{x} = \mathbf{b}'$ have the same sign pattern as the solutions of $\mathbf{Ax} = \mathbf{b}$. If this is the case then the system $\mathbf{Ax} = \mathbf{b}$ is called **sign solvable**.

Not surprisingly, the problem of characterizing the sign solvable systems is closely related to (in fact, equivalent to, as shown by Klee *et al.* [**31**]) the problem of characterizing the **sign non-singular** matrices. These matrices, also called *L-matrices*, are defined as follows: a matrix $\mathbf{B}$ is an **L-matrix** if, for any matrix $\mathbf{B}'$ with the same sign pattern as $\mathbf{B}$, the columns of $\mathbf{B}'$ are linearly independent. Klee *et al.* [**31**] showed that the problem of characterizing L-matrices is NP-complete, but they left open the important special case of characterizing *square* L-matrices.

Consider the following two matrices:

$$\begin{pmatrix} 1 & 1 & 1 \\ -1 & 1 & 1 \\ 0 & -1 & 1 \end{pmatrix}, \begin{pmatrix} 1 & 1 & 1 \\ -1 & 1 & 1 \\ 1 & -1 & 1 \end{pmatrix}.$$

In the first of these matrices, all terms in the standard expression of the determinant are non-negative and not all zero, and hence the matrix is an L-matrix. In the second matrix, the standard expression of the determinant has both positive terms and a negative term, and it is easy to see, by a continuity argument, that some matrix with the same sign pattern has vanishing determinant. So the second matrix is not an L-matrix.

The standard expression of a square L-matrix $\mathbf{B}$ must have a non-zero term. This corresponds to a perfect matching in the bipartite graph $G_{\mathbf{B}}$, whose partite sets correspond to the rows and columns of $\mathbf{B}$, and whose edges correspond to the non-zero entries of $\mathbf{B}$. Whether or not $G_{\mathbf{B}}$ has a perfect matching can be determined by a polynomially bounded algorithm (see [**10**]), and if $G_{\mathbf{B}}$ has a perfect matching, then we can permute rows and columns of $\mathbf{B}$ so that we get a matrix with no zeros on the main diagonal. Now these operations, together with multiplication of a column by a non-zero constant and multiplication of an entry of $\mathbf{B}$ by a positive constant, do not destroy the property of being an L-matrix. We conclude that, in order to investigate L-matrices, it is sufficient to consider matrices with entries in $\{0, 1, -1\}$ and with 1s on the main diagonal, as in the matrices above. Given such an $n \times n$ matrix $\mathbf{B} = (b_{ij})$, we form a digraph $D_{\mathbf{B}}$ with vertex-set $\{v_1, \ldots, v_n\}$ and with all arcs of the form v_iv_j whenever $b_{ij} \neq 0$. Furthermore, we assign the weight 1 to v_iv_j when $b_{ij} = 1$, and 0 when $b_{ij} = -1$. The digraphs in Fig. 4 correspond to the above matrices.

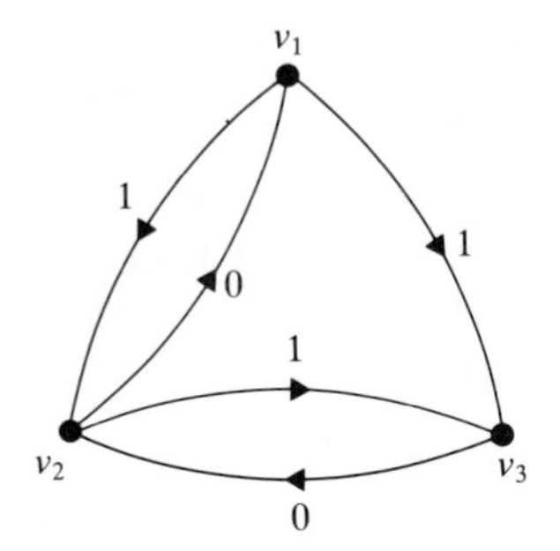

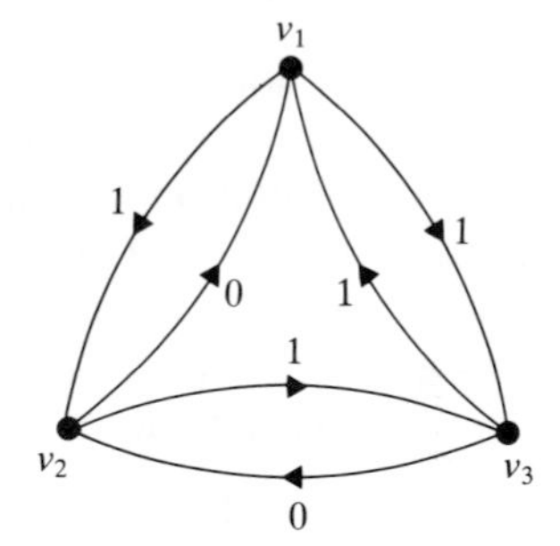

Fig. 4

Now each term in the standard expression of det $\mathbf{B}$ corresponds to a collection of disjoint directed circuits in $D_{\mathbf{B}}$. For example, when $\mathbf{B}$ is the second matrix above, the negative term corresponds to the directed circuit $v_1v_3v_1$ in Fig. 4. Note that this directed circuit has even total weight. This is easy to generalize as follows:

Theorem 8.1. *An $n \times n$ matrix $\mathbf{B}$ with entries in $\{0, 1, -1\}$ and with 1s on the main diagonal is an L-matrix if and only if $D_{\mathbf{B}}$ has no directed circuit of even weight.* ||

Finding a directed circuit of even weight is no more difficult than finding one of even length. Indeed, $D_{\mathbf{B}}$ has a directed circuit of even weight if and only if there is an even directed circuit in the digraph obtained from $D_{\mathbf{B}}$ by inserting a new vertex of in-degree and out-degree 1 on each arc of weight zero.

9. Even Directed Circuits and Color-critical Hypergraphs

Seymour [**47**] showed that even directed circuits play an important role in characterizing certain color-critical hypergraphs. While Seymour's paper [**47**] is concerned with hypergraphs, we extract in this section the graph theoretic part of it. A hypergraph $H = (V, E)$ is **2-colorable**, or **bipartite**, if there is a partition $V = V_1 \cup V_2$ such that each hyperedge intersects V_1

and V_2. H is **3-color-critical** if H is not bipartite, but every proper subhypergraph is.

Seymour's first elegant observation is the following:

Theorem 9.1. *If $H = (V, E)$ is a 3-color-critical hypergraph, then $|E| \geqslant |V|$.*

Proof. We regard each element of V as a real variable, and for each hyperedge A we consider the equation $\sum_{x \in A} x = 0$. Suppose $|E| < |V|$. Then there is a non-zero solution to the above system of equations. Let V_1 and V_2 be those elements in V which are, respectively, positive and negative in this solution, and put $V_3 = V \backslash (V_1 \cup V_2)$. Since $V_1 \cup V_2 \neq \emptyset$, the subhypergraph of H induced by V_3 is bipartite, and hence we can write $V_3 = V_1' \cup V_2'$, where each hyperedge in V_3 intersects both V_1' and V_2'. Now every hyperedge intersects both $V_1 \cup V_1'$ and $V_2 \cup V_2'$, giving the required contradiction. ||

The inequality of Theorem 9.1 is best possible, as shown in the next result. We need the following definitions: if D is a digraph and $v \in V(D)$, then $N(v)$ is the set consisting of v and the vertices dominated by v, and $H(D)$ is the hypergraph with vertex-set $V(D)$ and hyperedge-set $\{N(v) | v \in V(D)\}$.

Theorem 9.2. *If D is a strong digraph, then $H(D)$ is 3-color-critical if and only if D has no even directed circuit.*

Proof. Suppose first that D has an even directed circuit C. Since D is strongly connected, C can be extended to a spanning subdigraph D' of D, in such a way that the underlying graph of D' is connected and has only one circuit (namely C), and such that each arc in D' is directed towards C. This type of digraph is shown in Fig. 5.

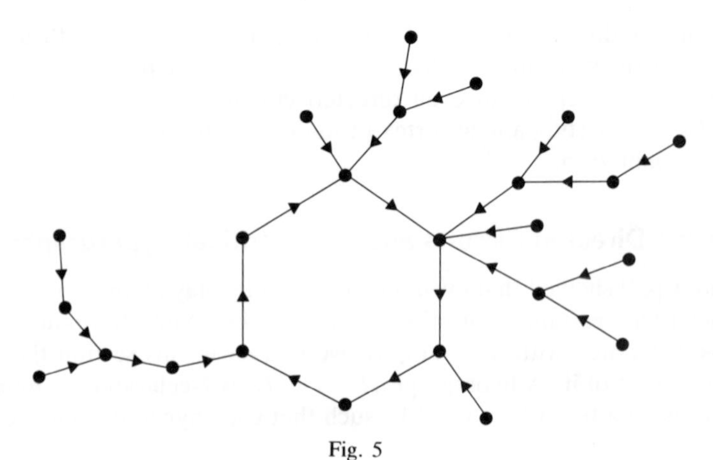

Fig. 5

Now D' is bipartite (in the graph sense), and any 2-coloring of D' is also a 2-coloring of H.

Suppose next that H has a 2-coloring. Then any vertex of D dominates a vertex of opposite color. This implies that there exists a directed circuit with alternating colors—that is, an even directed circuit.

We have shown that H is 2-colorable if and only if D has an even directed circuit. Suppose now that D has no even directed circuit, and let $N(v)$ be any hyperedge of D. Let T be a spanning tree of D, directed towards v. If we 2-color T (in the ordinary graph sense), then we obtain a 2-coloring of $H - N(v)$. This shows that H is 3-color-critical. ||

As we shall see in the next section, Theorem 9.2 gives a great variety of 3-color-critical hypergraphs with the same number of vertices and hyperedges. The next result (also due to Seymour [**47**]) shows that they all arise in that way:

Theorem 9.3. *If $H = (V, E)$ is a 3-color-critical hypergraph with $|V| = |E|$, then there exists a strongly connected digraph D with no even directed circuit, such that $H = H(D)$.*

Proof. We apply Hall's theorem to find a system of distinct representatives for the set of hyperedges of H. Suppose, to obtain a contradiction, that this is not so. Then, by Hall's theorem, H has a collection of hyperedges $A_1, \ldots, A_k$ such that $|A_1 \cup \ldots \cup A_k| < k$. We suppose that k is maximum. Since $|V| = |E|$, there must be hyperedges distinct from $A_1, \ldots, A_k$, and since k is maximum, no such hyperedge B is contained in $A = A_1 \cup \ldots \cup A_k$. Again, the maximality of k implies that the collection of sets $B \backslash A$, where B is a hyperedge not in $\{A_1, \ldots, A_k\}$, has a system of distinct representatives. We now form the digraph D' with vertex-set V, such that each of the above representatives dominates each vertex in the hyperedge of H which it represents.

Clearly, each vertex of A constitutes a terminal component of D'. There may be other terminal components with just one vertex, but we claim that each terminal component of D' which contains arcs has an even directed circuit. For otherwise, those hyperedges of H which are in a terminal component of D' with no even directed circuit form a proper non-bipartite subhypergraph of H, which is a contradiction. Now select an even directed circuit in each terminal component of D' with more than one vertex. D' has a spanning subdigraph D'' such that each component in the underlying undirected graph is either a tree, which in D'' is directed towards a vertex which is a terminal component of D'', or else a digraph of the type shown in Fig. 5. Since H is 3-color-critical, there is a bipartition (that is, a 2-coloring) of A, such that each of $A_1, \ldots, A_k$ intersects each partite set. This 2-coloring can be extended to a 2-coloring of D' such that each arc of D' has ends of different colors. This gives a 2-coloring of H. For, if B is a hyperedge not in $\{A_1, \ldots, A_k\}$, then B has a representative (b, say) which is not in A. In D', b has out-degree at least 1, since a 3-color-critical hypergraph with more than one vertex has no hyperedge with just one vertex. So b is not a terminal component of D', and hence b dominates a vertex b' in D''. This vertex b' has a different color from that of b, and hence B is 2-colored. But then H is bipartite, which is a contradiction.

We have shown that the hyperedges of H have a system of distinct representatives. Let D be the digraph obtained by letting each representative dominate each other vertex in the hyperedge of H which it represents. Then $H = H(D)$. If each terminal component of D has an even directed circuit, then we conclude as above that H is bipartite. We can therefore assume that some terminal component D''' has no even directed circuit. By Theorem 9.2, those hyperedges of H which lie in D''' constitute a 3-color-critical hypergraph. Thus $D''' = D$, and the proof is complete. ||

10. Degree Conditions for Even Directed Circuits

Theorems 8.1 and 9.3 suggest the problems of characterizing those digraphs with no even directed circuit, and of describing a good algorithm for finding an even directed circuit if it is present. These problems are unsolved.

One possible way of attacking them would be to find an even directed circuit passing through a prescribed arc. However, that problem is *NP*-complete, as shown by the next result of Fortune, Hopcroft and Wyllie **[18]**; this is in sharp contrast to Theorem 3.5 and the result in **[45]**, **[48]**, **[53]**:

Theorem 10.1. *It is NP-complete to find two disjoint directed paths with prescribed ends in a digraph.* ||

Corollary 10.2. *It is NP-complete to find an even directed circuit through a prescribed arc in a digraph.* ||

Proof. We reduce the problem in Theorem 10.1 to the problem in Corollary 10.2. Let D be a digraph with vertices v_1, v_2, w_1, w_2. We seek two disjoint dipaths from v_1 to w_1, and from v_2 to w_2, respectively. We now subdivide each arc of D once, and we add the arcs w_2v_1, w_1v_2. The resulting digraph has an even directed circuit through w_2v_1 if and only if D has two disjoint directed paths from v_1 to w_1 and from v_2 to w_2. ||

By the same reasoning we conclude the following result:

Corollary 10.3. *For each fixed natural number k ($k \geq 3$), it is NP-complete to find a dicycle of length 2 (modulo k) in a digraph.* ||

As pointed out in [**18**], Theorem 10.1 also implies the following; in this result, K_3^* is the complete symmetric digraph on three vertices:

Corollary 10.4. *It is NP-complete to find a subdivision of K_3^*.* ||

Motivated by Theorem 9.3, Lovász [**37**] and Seymour [**46**] asked whether every digraph of minimum out-degree at least 10^{10} contains an even directed circuit. This was answered in the negative in [**62**]:

Theorem 10.5. *For each natural number k, there exists a digraph D_k with minimum out-degree $k + 1$, and no even directed circuit.*

Proof. We let D_1 be the dicycle of length 3. Suppose that D_k exists, and consider a vertex v in D_k. We add to D_k $k + 2$ new vertices $v_1, \ldots, v_{k+2}$, where $v_1, \ldots, v_{k+1}$ dominate v and all the vertices of D_k which are dominated by v. We let v_{k+2} dominate $v_1, \ldots, v_{k+1}$, and finally we let v dominate v_{k+2}. The resulting digraph, which is shown in Fig. 6, has no even directed circuit.

If we perform this operation for each vertex v in D_k, we obtain a digraph D_{k+1} with no even dicircuit and minimum out-degree $k + 1$. ||

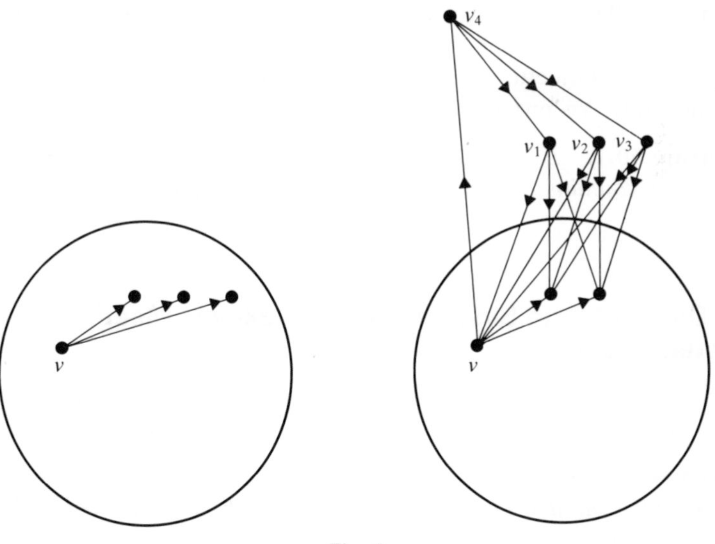

Fig. 6

In [**62**], Theorem 10.5 was refined as follows:

Theorem 10.6. (i) *If D is a digraph with p vertices and with minimum* *out-degree* $\lfloor \log_2 p \rfloor + 1$, *and if a weight (zero or one) is assigned to each arc, then D has a directed circuit of even weight;*

(*ii*) *for each natural number* $p \geq 2$, *there exists an arc-weighted digraph with p vertices, with minimum out-degree* $\lfloor \log_2 p \rfloor$, *and with no directed circuit of even weight;*

(*iii*) *for each natural number* $p \geq 2$, *there exists a digraph with p vertices, with minimum out-degree* $\lfloor \frac{1}{2} \log_2 p \rfloor$, *and with no even directed circuit.* ||

Combined with Theorem 9.3, Theorem 10.6 shows that there exist 3-color-critical hypergraphs with p vertices and p hyperedges each of which has cardinality at least $\lfloor \frac{1}{2} \log_2 p \rfloor$. On the other hand, there must be a hyperedge of cardinality at most $\lfloor \log_2 p \rfloor$. By Theorem 8.1, we obtain a similar extremal result on the minimum number of non-zero entries in the rows of an $n \times n$ L-matrix. By different graph-theoretic methods, a related result was obtained in [**64**]:

Theorem 10.7. *An* $n \times n$ *L-matrix has at least* $\frac{1}{2}(n-1)(n-2)$ *zero entries.* ||

The bound $\frac{1}{2}(n-1)(n-2)$ is sharp, as shown by the L-matrices which have 1 on and above the main diagonal, and -1 just below the main diagonal, as the first matrix in Section 8. All the extremal L-matrices are characterized in graph-theoretic terms in [**64**].

Lovász [**37**] conjectured that every 10^{10}-connected digraph contains an even dicircuit. We propose a stronger conjecture:

Conjecture 10.1. *Every* 10^{10}*-connected digraph whose arcs are weighted with* 0 *or* 1 *contains a directed circuit of even weight.*

11. Directed Path Systems and Vertex-sets Meeting all Directed Circuits

Fortune, Hopcroft and Wyllie [**18**] proved the following positive result for acyclic digraphs:

Theorem 11.1. *Let k be a fixed natural number. Then there exists a polynomially bounded algorithm for finding k disjoint directed paths with prescribed ends in an acyclic digraph.*

Proof. Suppose that we want to find k disjoint directed paths from v_i to w_i, for $i = 1, \ldots, k$, in an acyclic digraph D. We let D' denote the digraph whose vertices are all k-tuples of distinct vertices of D. Consider such a k-tuple $(z_1, \ldots, z_k)$. Among $z_1, \ldots, z_k$, there is one (z_j, say), which cannot be reached by any of the other z_i. If z_j dominates z_j', then we say that $(z_1, \ldots, z_j, \ldots, z_k)$ dominates $(z_1, \ldots, z_j', \ldots, z_k)$ in D'. Now if D' has a directed path from $(v_1, \ldots, v_k)$ to $(w_1, \ldots, w_k)$, then the vertices of the ith entries form a directed path from v_i to w_i, for $i = 1, \ldots, k$. These directed paths are pairwise disjoint, because of the way in which we have defined dominance in D'.

Conversely, if D has the k desired directed paths, then it is easy to use these to produce a directed path in D' from $(v_1, \ldots, v_k)$ to $(w_1, \ldots, w_k)$. Since D' can be constructed from D in polynomial time, and since a directed path from one vertex to another can be found in polynomial time (for example by Dijkstra's algorithm—see **[10]**), we have completed the proof. ||

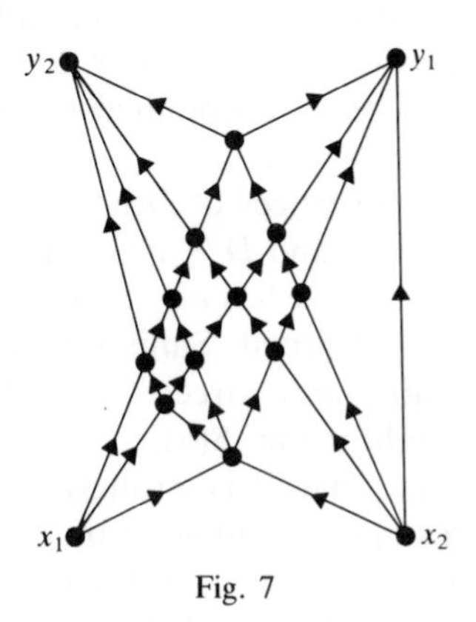

Fig. 7

A detailed investigation of the case $k = 2$ of Theorem 11.1 may be found in **[63]**. Figure 7 shows a planar acyclic digraph with no two disjoint directed paths from x_1 to y_1 and from x_2 to y_2, respectively. In **[63]** it was shown that if D is an acyclic digraph with no two disjoint directed paths from x_1 to y_1 and from x_2 to y_2, respectively, then D is essentially of the form shown in Fig. 7. That result turns out to be useful for studying digraphs with no vertices that meet all directed circuits. It implies the following result:

Theorem 11.2. *A digraph D contains a vertex meeting all directed circuits (that is, a vertex v such that $D - v$ is acyclic) if and only if it does not contain a subdivision of any of the digraphs of Fig. 8.*

Proof. Clearly, no digraph which contains a subdivision of a digraph in Fig. 8 has a vertex that meets all directed circuits. We prove, by induction on the number of vertices, that the converse holds.

So assume that $D - v$ contains a directed circuit for every vertex v in D. If $|V(D)| \leq 3$, then D is the digraph of Fig. 8(b), so we may assume

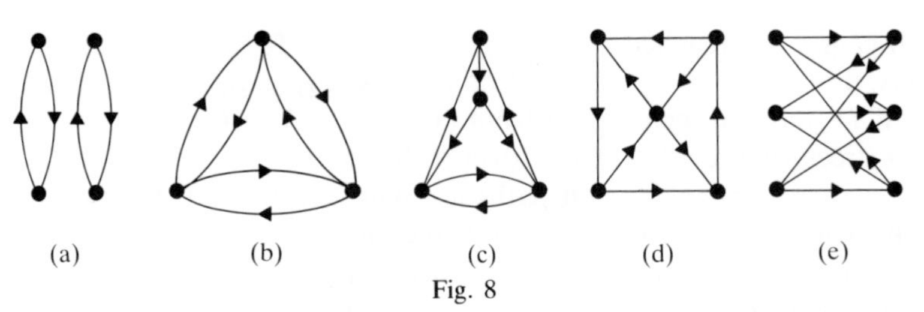

Fig. 8

that $|V(D)| \geq 4$. If D has a vertex with in-degree or out-degree zero, we delete it and use induction. If D has an arc vw such that either v has out-degree 1 or w has in-degree 1, then we contract the arc vw. It is easy to see that the resulting digraph has no vertex that meets all directed circuits, and hence it contains a subdivision of some digraph of Fig. 8. But then D also contains such a subdivision. (In order to see this, note that the digraphs of Figs. 8(b), (c), (d) are obtained by contracting one, two or three horizontal arcs in Fig. 8(e).)

So we can assume that all vertices of D have in-degree and out-degree at least 2. If v is a vertex of D and $D - v$ is not strongly connected, then any terminal component of $D - v$ has minimum out-degree at least 1, and therefore contains a directed circuit. Similarly, any strong component of $D - v$ with no incoming arcs has a directed circuit, and hence D contains a subdivision of the digraph of Fig. 8(a).

We can therefore assume that D is strongly 2-connected. We select a vertex v, and two vertices v_1, v_2 dominated by v, and two vertices w_1, w_2 dominating v. By Menger's theorem, D has two disjoint directed paths from $\{v_1, v_2\}$ to $\{w_1, w_2\}$, and hence D has two directed circuits C_1 and C_2 which have only v in common. Now $D - v$ has a directed circuit C_3, and it is an easy exercise to show that $C_1 \cup C_2 \cup C_3$ contains a subdivision of one of the digraphs of Figs. 8(a), (b), (c), (d). ||

Any subdivision of a digraph D in Fig. 8 contains a directed circuit of length at most $\frac{1}{3}|V(D)|$ (and not necessarily one of smaller length). We therefore get the following result of Allender [**3**]:

Corollary 11.3. *Let D be a digraph with p vertices. If D has (directed) girth greater than $\frac{1}{3}p$, then it has a vertex meeting all directed circuits.* ||

We also get the following Helly-type result of Kosaraju [**32**]:

Corollary 11.4. *If D is a digraph such that any three directed circuits have a vertex in common, then all directed circuits have a common vertex.*

Proof. If there is no vertex meeting all directed circuits, then D contains a subdivision of one of the digraphs of Fig. 8. But then it is easy to find three directed circuits with empty intersection. ||

Since any subdivision of a digraph in Figs. 8(b), (c), (d), (e) contains an even directed circuit, we also get the following result:

Corollary 11.5. *If a digraph D has no vertex meeting all directed circuits, then D contains either two disjoint directed circuits or an even directed circuit.* ||

In **[49]**, there is a precise description of all digraphs D with the property that any subdivision of D contains an even directed circuit. A refinement and further applications of Theorem 11.2 are given in **[65]**.

It was pointed out in **[64]** that if a digraph D contains a set A of k vertices (for k fixed) such that $D - A$ is acyclic, then the directed circuit length distribution can be determined in polynomial time. The same argument shows that the maximum number of pairwise disjoint directed circuits can be found in polynomial time. Thus Corollary 11.5 gives a polynomially bounded algorithm for finding either two disjoint directed circuits or an even directed circuit. We have already mentioned as unsolved the problem of finding an even directed circuit. If true, the following conjecture made by Gallai for $k = 2$, and generalized by Younger (see **[5]**, **[10]**), would give a polynomially bounded algorithm for finding k disjoint directed circuits (where k is fixed):

Conjecture 11.1. *For each natural number k, there exists a natural number $\pi(k)$ such that every digraph D containing no k disjoint directed circuits has a set A of at most $\pi(k)$ vertices such that $D - A$ is acyclic.*

A related statement was proved in **[58]**:

Theorem 11.6. *For each natural number k, there exists a natural number $\mu(k)$ such that every digraph with minimum out-degree at least $\mu(k)$ contains k disjoint directed circuits.*

D. H. Younger **[72]** has proved that $\pi(2) \geq 3$, and it was shown in **[58]** that $\mu(2) = 3$.

12. Subdivisions in Digraphs and Tournaments

By the remark preceding Corollary 11.5, any subdivision of K_3^* contains an even directed circuit, and it follows from Theorem 10.5 that a large (fixed) minimum out-degree in a digraph does not imply the presence of a subdivision of K_3^*. It was shown in **[63]** that every digraph with minimum out-degree at least 2 contains two directed circuits with precisely one vertex in common, but examples in **[62]** show that a large (fixed) minimum out-degree does not imply the existence of three directed circuits with precisely one vertex in common, pair by pair. So if we seek directed analogues of Theorem 2.3, we should confine ourselves to highly disconnected digraphs, as in Theorem 11.6, or acyclic digraphs, as suggested by the following problem of Mader [**41**]:

Conjecture 12.1. *For each natural number k, there exists a natural number $h'(k)$ such that every digraph with minimum out-degree at least $h'(k)$ contains a subdivision of the transitive tournament on k vertices.*

Let m be a natural number. Since any undirected graph with minimum degree at least $2m$ has an orientation with minimum in-degree and out-degree at least m, it follows that

$$h(k) \leq 2h'(k),$$

where $h(k)$ is as defined after Theorem 2.3.

Another possible approach to subdivisions in digraphs is an analogue of Corollary 3.2 (see [**62**]):

Conjecture 12.2. *For each natural number k, there exists a natural number $g'(k)$ such that, for any distinct vertices $v_1, \ldots, v_k, w_1, \ldots, w_k$ in a strongly $g'(k)$-connected digraph D, there are k disjoint directed paths $P_1, \ldots, P_k$ in D such that P_i starts at v_i and ends at w_i, for $i = 1, \ldots, k$.*

If true, Conjecture 12.2 implies the existence of subdivisions of large complete digraphs in digraphs with large strong connectivity. In particular, Conjecture 12.2 implies Conjecture 10.1, except that 10^{10} is replaced by $g'(6) + 3$.

Finally, some of the results on undirected graphs discussed in this chapter have natural extensions for tournaments, as shown in [**61**]. We conclude by discussing some of these results, beginning with an analogue of Conjecture 6.4 for tournaments:

Theorem 12.1. *Let v and w be vertices in a $(k + 4)$-connected tournament T, where $k \geq 1$, and let P be a shortest vw directed path. Then $T - V(P)$ is k-connected.* ||

The ideas of the proof of Theorem 12.1 were used to give the following analogue of Conjecture 12.2 and Corollary 3.2:

Theorem 12.2. *For each natural number k, there exists a natural number $g''(k)$ such that the following holds: if T is a tournament, and $v_1, \ldots, v_k, w_1, \ldots, w_k$ are distinct vertices in T such that, for each $i = 1, 2, \ldots, k$, there are $g''(k)$ internally disjoint directed paths from v_i to w_i, for $i = 1, \ldots, k$, then T has k disjoint directed paths $P_1, \ldots, P_k$ such that P_i starts at v_i and ends at w_i, for $i = 1, 2, \ldots, k$.* ||

Note that Theorem 12.2 involves only local strong connectivity. If we impose a global strong connectivity condition on the tournament, we can do even better:

Theorem 12.3. *If* $v_1, \ldots, v_k, w_1, \ldots, w_k$ *are distinct vertices in a strongly* $(g''(5k) + 12k + 9)$*-connected tournament* T, *then* T *has* k *disjoint directed paths* $P_1, \ldots, P_k$ *such that*:

(*i*) P_i *starts at* x_i *and ends at* y_i, *for* $i = 1, \ldots, k$, *and*

(*ii*) $V(P_1) \cup \ldots \cup V(P_k) = V(T)$. ||

Even the case $k = 1$ in Theorem 12.3 is difficult. It was shown in **[52]** that every 4-connected tournament has a Hamiltonian directed path with prescribed starting and ending vertices.

Theorem 12.3 implies that any tournament with large strong connectivity contains a spanning subdigraph which is a subdivision of K_k^*.

The connectivity results for undirected graphs suggest a number of problems concerning tournaments. We mention here just one such problem, motivated by Theorem 6.6:

Conjecture 12.3. *For each natural number* k, *there exists a natural number* $v(k)$ *such that the vertex-set of any strongly* $v(k)$*-connected tournament can be partitioned into two sets, each of which induces a strongly* k*-connected subtournament.*

References

1. A. V. Aho, J. E. Hopcroft and J. D. Ullman, *The Design and Analysis of Computer Algorithms*, Addison-Wesley, Reading, Mass., 1975; *MR***54**#1706.
2. R. E. L. Aldred, D. A. Holton and C. Thomassen, Cycles through four edges in 3-connected cubic graphs, *Graphs and Combinatorics* **1** (1985), 7–11; *MR***86i**:05092.
3. E. Allender, On the number of cycles possible in digraphs with large girth, *Discrete Appl. Math.* **10** (1985), 211–225; *MR***86e**:05043.
4. N. Alon, D. Kleitman, M. Saks, P. D. Seymour and C. Thomassen, Subgraphs of large connectivity and chromatic number in graphs of large chromatic number, *J. Graph Theory* **11** (1987), 367–371.
5. J.-C. Bermond and C. Thomassen, Cycles in digraphs—a survey, *J. Graph Theory* **5** (1981), 1–43; *MR***82k**:05053.
6. B. Bollobás, Cycles modulo k, *Bull. London Math. Soc.* **9** (1977), 97–98; *MR***56**#181.
7. B. Bollobás, Semi-topological subgraphs, *Discrete Math.* **20** (1977), 83–85; *MR***80a**:05125.
8. B. Bollobás, *Extremal Graph Theory*, Academic Press, London, 1978; *MR***80a**:05120.
9. B. Bollobás, Cycles and semi-topological configurations, *Theory and Applications of Graphs* (ed. Y. Alavi and D. R. Lick), Lecture Notes in Math. **642**, Springer, Berlin, 1978, pp. 66–74; *MR***80a**:05077.
10. J. A. Bondy and U. S. R. Murty, *Graph Theory with Applications*, American Elsevier, New York, and MacMillan, London, 1976; *MR***54**#117.
11. P. A. Catlin, Hajós' graph-coloring conjecture: variations and counter-examples, *J. Combinatorial Theory (B)* **26** (1979), 268–274; *MR***81g**:05057.
12. K. Corradi and A. Hajnal, On the maximal number of independent circuits of a graph, *Acta Math. Acad. Sci. Hungar.* **14** (1963), 423–439; *MR***34**#84.

13. G. A. Dirac, In abstrakten Graphen vorhandene vollständige 4-Graphen und ihre Unterteilungen, *Math. Nachr.* **22** (1960), 61–85; *MR*22#12053.
14. G. A. Dirac, Homomorphism theorems for graphs, *Math. Ann.* **153** (1964), 69–80; *MR*28#3417.
15. G. A. Dirac, Chromatic number and topological complete subgraphs, *Canad. Math. Bull.* **8** (1965), 711–715; *MR*33#3955.
16. P. Erdős, Graph theory and probability, *Canad. J. Math.* **11** (1959), 34–38; *MR*21#876.
17. P. Erdős and S. Fajtlowicz, On a conjecture of Hajós, *Combinatorica* **1** (1981), 141–143; *MR*83d:05042.
18. S. Fortune, J. Hopcroft and J. Wyllie, The directed subgraph homeomorphism problem, *J. Theoret. Comput. Sci.* **10** (1980), 111–121; *MR*81e:68079.
19. D. Grant, F. Jaeger and C. Payan, On graphs without antidirected elementary cycles and related topics, preprint, 1979.
20. R. L. Graham, On subtrees of directed graphs with no path of length exceeding one, *Canad. Math. Bull.* **13** (1970), 329–332; *MR*42#5859.
21. A. Gyárfás, J. Komlós and E. Szemerédi, On the distribution of cycle lengths in graphs, *J. Graph Theory* **4** (1984), 441–462; *MR*85k:05068.
22. A. Gyárfás, H. J. Prömel, E. Szemerédi and B. Voigt, On the sum of the reciprocals of cycle lengths in sparse graphs, *Combinatorica* **5** (1985), 41–52; *MR*86m:05059.
23. P. Hajnal, Partition of graphs with condition on the connectivity and minimum degree, *Combinatorica* **3** (1983), 95–99; *MR*85b:05118.
24. R. Häggkvist, Equicardinal disjoint cycles in sparse graphs, *Ann. Discrete Math.* **27** (1985), 269–273.
25. R. Häggkvist and C. Thomassen, Circuits through specified edges, *Discrete Math.* **41** (1982), 29–34; *MR*84g:05095.
26. D. A. Holton and C. Thomassen, Research problem, *Discrete Math.*, to appear.
27. H. A. Jung, Anwendung einer Methode von K. Wagner bei Färbungsproblemen für Graphen, *Math. Ann.* **161** (1965), 325–326; *MR*32#5542.
28. H. A. Jung, Eine Verallgemeinerung des n-fachen Zusammenhangs für Graphen, *Math. Ann.* **187** (1970), 95–103; *MR*42#2966.
29. R. M. Karp, On the complexity of combinatorial problems, *Networks* **5** (1975), 45–68.
30. A. K. Kelmans, A strengthening of the Kuratowski planarity criterion for 3-connected graphs, *Discrete Math.* **51** (1984), 215-220; *MR*86j:05054.
31. V. Klee, R. Ladner and R. Manber, Signsolvability revisited, *Linear Algebra and Applications* **59** (1984), 131–157.
32. S. R. Kosaraju, On independent circuits of a digraph, *J. Graph Theory* **1** (1977) 379–382; *MR*58#346.
33. A. V. Kostochka, Lower bound of the Hadwiger number of graphs by their average degree, *Combinatorica* **4** (1984), 307–316; *MR*86k:05067.
34. D. G. Larman and P. Mani, On the existence of certain configurations within graphs and the 1-skeletons of polytopes, *Proc. London Math. Soc.* **20** (1970), 144–160; *MR*41#8288.
35. L. Lovász, On decomposition of graphs, *Studia Sci. Math. Hungar.* **1** (1966), 237–238; *MR*34#2492.
36. L. Lovász, Problem 5, *Period. Math. Hungar.* **4** (1974), 82.
37. L. Lovász, Problems in *Recent Advances in Graph Theory* (ed. M. Fiedler), Academia, Prague, 1975.
38. W. Mader, Homomorphieeigenschaften und mittlere Kantendichte von Graphen, *Math. Ann.* **174** (1967), 265–268; *MR*36#3668.
39. W. Mader, Existenz n-fach zusammenhängenden Teilgraphen in Graphen genügend grosser Kantendichte, *Abh. Math. Sem. Univ. Hamburg* **37** (1972), 86–97; *MR*46#5177.
40. W. Mader, Hinreichende Bedingungen für die Existenz von Teilgraphen, die zu einem vollständigen Graphen homöomorph sind, *Math. Nachr.* **53** (1972), 145–150; *MR*58#16421.
41. W. Mader, Degree and local connectivity in digraphs, *Combinatorica* **5** (1985), 161–165.
42. J. Pelikán, Valency conditions for the existence of certain subgraphs, *Theory of Graphs* (ed. P. Erdős and G. Katona), Academic Press, New York, 1968, pp. 251–258; *MR*38#3171.

43. N. Robertson and P. D. Seymour, Graph minors—a survey, preprint, Bell Comm. Res., 1985.
44. P. A. Samuelson, *Foundations of Economic Analysis*, Atheneum, New York 1971; originally published by Harvard University Press, 1947.
45. Y. Shiloach, A polynomial solution to the undirected two paths problem, *J. Assoc. Comput. Math.* **27** (1980), 445–456; *MR***82a**:05061.
46. P. D. Seymour, D. Phil. Thesis, Oxford University, 1973.
47. P. D. Seymour, On the two-colouring of hypergraphs, *Quart. J. Math. (Oxford)* **25** (1974), 303–312; *MR***51**#7927.
48. P. D. Seymour, Disjoint paths in graphs, *Discrete Math.* **29** (1980), 293–309; *MR***82b**:05091.
49. P. D. Seymour and C. Thomassen, Characterization of even digraphs, *J. Combinatorial Theory* (*B*) **42** (1987), 36–45.
50. C. Thomassen, Some homeomorphism properties of graphs, *Math. Nachr.* **64** (1974), 119–133; *MR***51**#288.
51. C. Thomassen, A minimal condition implying a special K_4 subdivision in a graph, *Arch. Math.* **25** (1974), 210–215; *MR***50**#12811.
52. C. Thomassen, Hamiltonian-connected tournaments, *J. Combinatorial Theory (B)* **28** (1980), 142–163; *MR***82d**: 05065.
53. C. Thomassen, 2-linked graphs, *Europ. J. Combinatorics* **1** (1980), 371–378; *MR***82c**:05086.
54. C. Thomassen, Nonseparating cycles in *k*-connected graphs, *J. Graph Theory* **5** (1981), 351–354; *MR***83f**:05038.
55. C. Thomassen, Graph decomposition with constraints on the connectivity and minimum degree, *J. Graph Theory* **7** (1983), 165–167; *MR***84i**:05091.
56. C. Thomassen, Graph decomposition with applications to subdivisions and path systems modulo *k*, *J. Graph Theory* **7** (1983), 261–271; *MR***85e**:05106.
57. C. Thomassen, Girth in graphs, *J. Combinatorial Theory (B)* **35** (1983), 129–141; *MR***85f**:05076.
58. C. Thomassen, Disjoint cycles in digraphs, *Combinatorica* **3** (1983), 393–396; *MR***85e**:05087.
59. C. Thomassen, Subdivisions of graphs with large minimum degree, *J. Graph Theory* **8** (1984), 23–28; *MR***85k**: 05097.
60. C. Thomassen, A refinement of Kuratowski's theorem, *J. Combinatorial Theory (B)* **37** (1984), 245–253; *MR***86e**:05059.
61. C. Thomassen, Connectivity in tournaments, *Graph Theory and Combinatorics* (ed. B. Bollobás), Academic Press, London, 1984, pp. 305–313; *MR***86g**:05062.
62. C. Thomassen, Even cycles in directed graphs, *Europ. J. Combinatorics* **6** (1985), 85–89; *MR***86i**:05098.
63. C. Thomassen, The 2-linkage problem for acyclic digraphs, *Discrete Math.* **55** (1985), 73–87; *MR***86k**:05078.
64. C. Thomassen, Sign-nonsingular matrices and even cycles in directed graphs, *Linear Algebra and Applications* **75** (1986), 27–41.
65. C. Thomassen, On directed graphs with no two disjoint directed cycles, *Combinatorica* **7** (1987), 145–150.
66. C. Thomassen, Configurations in graphs of large minimum degree, connectivity or chromatic number, to appear.
67. C. Thomassen and B. Toft, Induced nonseparating cycles in graphs, *J. Combinatorial Theory (B)* **31** (1981), 199–224; *MR***82m**:05062.
68. B. Toft, Problem 10, *Recent Advances in Graph Theory* (ed. M. Fiedler), Academia, Prague, 1975, p. 544.
69. W. T. Tutte, How to draw a graph, *Proc. London Math. Soc.* **13** (1963), 743–767; *MR***28**#1610.
70. K. Wagner, Beweis einer Abschwächung der Hadwiger-Vermutung, *Math. Ann.* **153** (1964), 139–141; *MR***28**#3416.
71. D. R. Woodall, Circuits containing specified edges, *J. Combinatorial Theory (B)* **22** (1977), 274–278; *MR***55**#12572.
72. D. H. Younger, Graphs with interlinked directed circuits, *Proc. Midwest Symp. on Circuit Theory* **2** (1973), XVI 2.1–XVI 2.7.

6
Isometric Embeddings of Graphs

R. L. GRAHAM

1. Introduction

With any connected graph $G = (V, E)$ one can associate a metric d_G: $V \times V \to \mathbf{N}$ (the set of non-negative integers) by defining $d_G(v, w)$, for $v, w \in V$, to be the minimum number of edges in any path between v and w. This is the most common definition of *distance* in a graph and has been investigated by many researchers over the years.

In this chapter we combine the basic concepts of distance and subgraph. More precisely, we say that G' is an **isometric** (or **distance-preserving**) **subgraph** of G if, for all vertices v and w, $d_{G'}(v, w) = d_G(v, w)$. Note that this is a natural strengthening of the concept of induced subgraph, since G' is an induced subgraph of G if, for all vertices v and w in V, $d_{G'}(v, w) = 1$ if and only if $d_G(v, w) = 1$.

We shall see that the requirements for a subgraph to be isometric are rather restrictive, and, consequently, a number of surprisingly strong conclusions can be deduced in this case.

2. A General Formulation

Suppose that (M, d) is a metric space—that is, M is a set and d is a mapping from $M \times M$ into $\mathbf{R}$ (the set of real numbers) satisfying, for all $x, y, z \in M$:

GRAPH THEORY, 3
ISBN 0-12-086203-4

(*i*) $d(x, y) = d(y, x) \geq 0$, with equality if and only if $x = y$;

(*ii*) $d(x, y) + d(y, z) \geq d(x, z)$ (the triangle inequality).

A mapping $\lambda: V \to M$ is said to be an **isometric embedding** of $G = (V, E)$ into M if

$$d(\lambda(v), \lambda(w)) = d_G(v, w),$$

for all $v, w \in V$. We will often abbreviate this by writing $\lambda: G \xrightarrow{I} V$, or even $G \xrightarrow{I} M$ if we mean that a suitable λ exists.

In Fig. 1(a) we show an example of an isometric embedding of C_6 into the cube Q_3, and in Fig. 1(b) we show an embedding of C_6 into Q_3 in which C_6 is an induced subgraph of Q_3 but the embedding is not isometric.

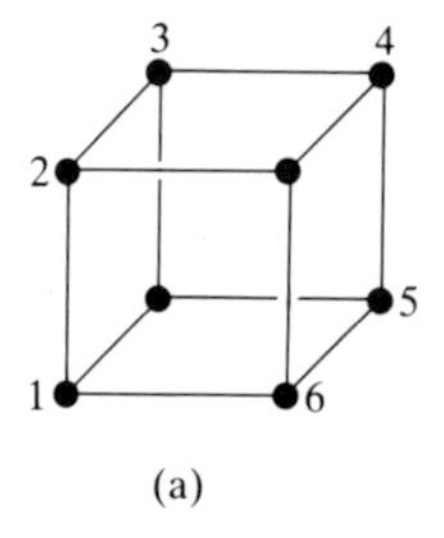

(a)

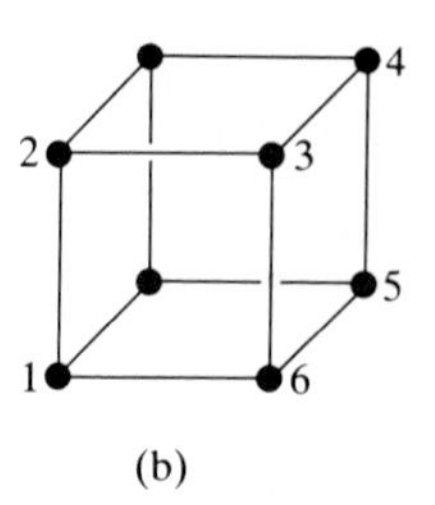

(b)

Fig. 1

Many of the spaces (M, d) we shall be concerned with are *product spaces*—that is, spaces formed as Cartesian products of smaller spaces with a much simpler distance structure. Specifically, if (M_k, d_k) $(1 \leq k \leq r)$ are metric spaces, then the product space (M^*, d^*) is defined by

$$M^* = \prod_{k=1}^{r} M_k = \{(m_1, \ldots, m_r): m_k \in M_k, \quad 1 \leq k \leq r\},$$

and

$$d^*((m_1, \ldots, m_r), (m'_1, \ldots, m'_r)) = \sum_{k=1}^{r} d_k(m_k, m'_k).$$

For example, if each M_k consists of the 2-point space $\{0, 1\}$ in which the distance between the two points 0 and 1 is 1, then (M^*, d^*) is just (Q_r, d_H), the r-cube Q_r equipped with the *Hamming metric* for which the distance between two binary r-triples is equal to the number of coordinate positions in which they differ.

3. Extended Binary Labelings

Our first set of results deals with one of the earliest developments of isometric embeddings of graphs. In this case, we wish to find efficient embeddings of graphs into B_*^r, where B_* consists of the set $\{0, 1, *\}$ with the distance d_* defined by

$$d_*(x, y) = \begin{cases} 1, & \text{if } x = 0,\ y = 1 \text{ or } x = 1,\ y = 0, \\ 0, & \text{otherwise.} \end{cases}$$

In other words, we wish to label each vertex v with an appropriate r-tuple $\lambda(v)$ so that the distances between vertices in V are exactly given by the distances between their corresponding labels. (This problem arose in connection with early work on routing algorithms for packet switching in data networks; see **[29]**, **[22]** and **[11]**.) In Fig. 2(a) we show a graph G and an appropriate labeling; in Fig. 2(b) we give its distance matrix $\mathbf{D}(G) = (d_{ij})$, where $d_{ij} = d_G(v_i, v_j)$.

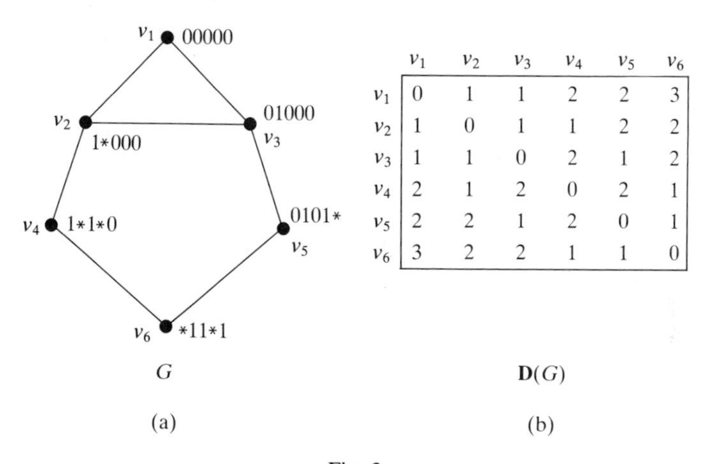

	v_1	v_2	v_3	v_4	v_5	v_6
v_1	0	1	1	2	2	3
v_2	1	0	1	1	2	2
v_3	1	1	0	2	1	2
v_4	2	1	2	0	2	1
v_5	2	2	1	2	0	1
v_6	3	2	2	1	1	0

D(G)

(a) (b)

Fig. 2

To begin with, it is not clear that such isometric embeddings always exist. For example, if we are not allowed to use the symbol $*$, then we are asking for isometric embeddings of G into $\{0, 1\}^r$ (that is, Q_r), and this is certainly *not* possible for most graphs (such as non-bipartite graphs). However, this is taken care of by the following result:

Lemma 3.1. *For each connected graph G there exists a least integer $r = r(G)$ such that $G \xrightarrow{I} B_*^r$.*

Proof. Let R denote the integer $\sum_{i<j} d_G(v_i, v_j) = \sum_{i<j} d_{ij}$. We claim that $G \xrightarrow{I} B_*^R$. To see this, for each i and j with $i < j$, select d_{ij} fixed (and mutually disjoint) coordinate positions $D_{ij} \subseteq \{1, 2, \ldots, R\}$ and define

$$\lambda(v)_k = \begin{cases} 0, & \text{if } v = v_i,\ k \in D_{ij}, \\ 1, & \text{if } v = v_j,\ k \in D_{ij}, \\ *, & \text{otherwise.} \end{cases}$$

Since $\{1, 2, \ldots, R\} = \bigcup_{i<j} D_{ij}$, an easy computation shows that λ: $G \xrightarrow{I} B_*^R$, and we are done. ||

An important question, open for many years, was *how large* $r(G)$ can ever be. This was finally settled by the following elegant result of Winkler **[36]**:

Theorem 3.2. $r(G) \leq |V(G)| - 1$. ||

This inequality improves earlier estimates of Yao **[37]** and others (see **[11]**, **[22]**), and is best possible for many infinite classes of graphs, as we shall see shortly.

In the other direction, the following lower bound for $r(G)$ was found by Witsenhausen (see **[22]**):

Theorem 3.3. $r(G) \geq \max\{n_+(G), n_-(G)\}$, *where* $n_+(G)$ *and* $n_-(G)$ *denote the number of positive and negative eigenvalues of the distance matrix of* G.

Proof. Suppose that λ: $G \xrightarrow{I} B_*^r$ is given. For $1 \leq k \leq r$, define subsets X_k and Y_k of $I = \{1, 2, \ldots, |V(G)|\}$ by $X_k = \{i \in I\colon \lambda(v_i)_k = 0\}$ and $Y_k = \{i \in I\colon \lambda(v_i)_k = 1\}$. From the hypothesis that λ is an isometry and the definition of d_*, we have

$$d_{ij} = d_G(v_i, v_j) = \sum_{k=1}^{r} d_*(\lambda(v_i)_k, \lambda(v_j)_k),$$

for all i and j. We can summarize this information in the following way:

$$\sum_{i<j} d_{ij} x_i x_j = \sum_{k=1}^{r} \left(\sum_{i \in X_k} x_i\right)\left(\sum_{j \in Y_k} x_j\right),$$

where the x_i are indeterminates. Thus, the existence of an isometric embedding of G into B_*^r implies the existence of a decomposition of the associated quadratic form $\sum_{i<j} d_{ij} x_i x_j$ into a sum of certain products. We

can go one step further and rewrite this equation as

$$\sum_{i<j} d_{ij}x_ix_j = \tfrac{1}{4}\sum_{k=1}^{r}\left\{\left(\sum_{i\in X_k} x_i + \sum_{j\in Y_k} x_j\right)^2 - \left(\sum_{i\in X_k} x_i - \sum_{j\in Y_k} x_j\right)^2\right\},$$

where we have simply used the algebraic identity

$$xy = \tfrac{1}{4}\left((x+y)^2 - (x-y)^2\right).$$

However, this is a decomposition of $\sum_{i<j} d_{ij}x_ix_j$ into a sum and difference of squares, to which we can apply Sylvester's classical law of inertia (see [**25**, p. 352]). This implies that the number of positive squares must be at least $n_+(G)$, and the number of negative squares must be at least $n_-(G)$. Since each of the numbers of positive and negative squares in the above decomposition is at most r, we have $r \geq n_+(G)$, $r \geq n_-(G)$, and the result follows. ||

Theorems 3.2 and 3.3 can be used to determine $r(G)$ exactly, for many graphs G. For example, if G is the complete graph K_p, then an easy computation shows that $n_-(K_p) = p - 1$, and consequently $r(K_p) = p - 1$. Similarly, for the odd circuit C_{2k+1}, we have

$$n_-(C_{2k+1}) = 2k = |V(C_{2k+1})| - 1,$$

and so $r(C_{2k+1}) = 2k$. In the case that G is a tree, much more can actually be said. This we do in the next section.

Before closing this section, we point out that the assertion $r(K_p) = p-1$ has the following equivalent combinatorial interpretation (see [**22**]):

Theorem 3.4. *It is not possible to decompose the edge-set of K_p into fewer than $p - 1$ edge-disjoint complete bipartite subgraphs.*

Remark. At present, no purely combinatorial proof of this is known. However, Tverberg [**32**] has given the following nice algebraic argument:

Proof. Suppose that $E(K_p) = \bigcup_{k=1}^{t} K(A_k, B_k)$ is a decomposition of the edge-set of K_p into t edge-disjoint complete bipartite subgraphs $K(A_k, B_k)$ $(1 \leq k \leq t)$. Then

$$\sum_{1\leq i<j\leq p} x_ix_j = \sum_{k=1}^{t}\left(\sum_{a\in A_k} x_a\right)\left(\sum_{b\in B_k} x_b\right).$$

Consider the following system of $t + 1$ homogeneous linear equations in the p variables x_i:

$$\sum_{a \in A_k} x_a = 0 \ (1 \leq k \leq t), \text{ and } \sum_{i=1}^{p} x_i = 0.$$

If $(y_1, \ldots, y_p)$ is any solution to this system, then we must have

$$\begin{aligned} 0 &= \left(\sum_{i=1}^{p} y_i\right)^2 = \sum_{i=1}^{p} y_i^2 + 2 \sum_{i<j} y_i y_j \\ &= \sum_{i=1}^{p} y_i^2 + 2 \sum_{k=1}^{t} \left(\sum_{a \in A_k} y_a\right)\left(\sum_{b \in B_k} y_b\right) \\ &= \sum_{i=1}^{p} y_i^2; \end{aligned}$$

that is, $y_i = 0$ for all i. Hence, the number of equations in the system must be as least as large as the number of variables, giving $t + 1 \geq n$. ||

4. Distance Matrices of Trees

Suppose that T_p is a tree with p vertices. An unexpected fact concerning the distance matrix $\mathbf{D}(T_p)$, first noted in **[22]**, is given by the following theorem. The main point of this result is that the value of the determinant is independent of the *structure* of T_p, and depends only on its order.

Theorem 4.1. $\det \mathbf{D}(T_p) = (-1)^{p-1}(p-1)2^{p-2}$.

Sketch of proof. We simply designate an arbitrary fixed vertex of T_p as a root, and sequentially perform row and column operations on $\mathbf{D}(T_p)$ by subtracting from the row and column of each vertex v the row and column of its immediate predecessor v' (so that v' is adjacent to v and lies on the path in T_p from v to the root), always processing the vertices furthest from the root first. When this is done, the resulting matrix $\mathbf{M} = (m_{ij})$ has the form

$$m_{ij} = \begin{cases} 1, & \text{if } i = 1 \text{ or } j = 1, \text{ but } (i, j) \neq (1, 1), \\ -2, & \text{if } i = j = 1, \\ 0, & \text{otherwise,} \end{cases}$$

and the result follows at once. ||

However, one suspects from the form of Theorem 4.1 that much more is going on here. This has led to a number of extensions which help shed light on its particularly simple form.

To state the first of these extensions, let cof $\mathbf{D}(H)$ denote the sum of the cofactors of the distance matrix $\mathbf{D}(H)$ of a graph H. The following result is given in **[20]**:

Theorem 4.2. *If a connected graph G has blocks $G_1, \ldots, G_r$, then*

(*i*) $\operatorname{cof} \mathbf{D}(G) = \prod_{k=1}^{r} \operatorname{cof} \mathbf{D}(G_k)$;

(*ii*) $\det \mathbf{D}(G) = \sum_{k=1}^{r} \det \mathbf{D}(G_k) \prod_{i \neq k} \operatorname{cof} \mathbf{D}(G_i)$. ||

Since any tree T_p with p vertices has exactly $p - 1$ blocks, each of which is a single edge K_2 with $\det \mathbf{D}(K_2) = -1$, $\operatorname{cof} \mathbf{D}(K_2) = 2$, Theorem 4.1 follows at once.

Note that, when $\operatorname{cof} \mathbf{D}(G) \neq 0$, then part (*ii*) of Theorem 4.2 can be written in the suggestive form

$$\frac{\det \mathbf{D}(G)}{\operatorname{cof} \mathbf{D}(G)} = \sum_{k=1}^{r} \frac{\det \mathbf{D}(G)}{\operatorname{cof} \mathbf{D}(G_k)}.$$

It is not difficult to see that, in fact, there always exist mappings $T_p \xrightarrow{I} B_*^{p-1}$ in which the symbol $*$ is never used. In other words, $T_p \xrightarrow{I} \{0, 1\}^{p-1}$.

The next generalization of Theorem 4.1 is the following:

Theorem 4.3. *Let* $\{\mathbf{a}_1, \mathbf{a}_2, \ldots, \mathbf{a}_p\} \subseteq \{0, 1\}^{p-1}$. *Then*

$$\det(d_H(\mathbf{a}_i, \mathbf{a}_j)) = (-1)^{p-1}(p - 1)2^{p-2}V^2,$$

where V denotes the p-dimensional volume of the parallelepiped spanned by the vectors $\mathbf{a}_k - \mathbf{a}_0$ $(1 \leq k \leq p)$. ||

The proof of this relies on the use of special determinants of the form $\det(\mathbf{x}_i \cdot \mathbf{x}_j)$, called *Gramians* (see [**18**, p. 250]), where $\mathbf{x}_i \cdot \mathbf{x}_j$ denotes the inner product of the vectors $\mathbf{x}_i$ and $\mathbf{x}_j$. The values of these determinants turn out to represent volumes of parallelepipeds in Euclidean space, and the factor 2^{p-1} arises directly from this interpretation.

The final remark we make concerning Theorem 4.1 is the following. Rather than just looking at the determinant of $\mathbf{D}(T_p)$, we could investigate the characteristic polynomial

$$\Delta_{T_p}(x) = \det(\mathbf{D}(T_p) - x\mathbf{I}) = \sum_{k=0}^{p} d_k(T_p)x^k.$$

It turns out (see [**21**], [**16**]) that each coefficient $d_k(T_p)$ represents a fixed *linear combination* (independent of T_p) of the number of occurrences of various subforests in T_p (multiplied by a factor of 2^{p-k-2}). For the case $k = 0$, the coefficient

$$d_0(T_p) = \det \mathbf{D}(T_p)$$

depends only on the number of edges in T_p. The coefficients of these linear combinations themselves satisfy some rather mysterious relations which are not yet completely understood. However, a fuller description of this intriguing subject is beyond the scope of the present discussion. We point out that a curious lemma needed in **[21]** is the following:

Lemma 4.5. *For a tree T_p on p vertices, let ρ_i denote the degree of vertex v_i in T_p, and let $a_{ij} = 1$ if $\{v_i, v_j\} \in E(T_p)$, and 0, otherwise. Then the inverse $\mathbf{D}^{-1}(T_p) = (d^*_{ij})$ of $\mathbf{D}(T_p)$ is given by:*

$$d^*_{ij} = \frac{(2-\rho_i)(2-\rho_j)}{2(p-1)} + \begin{cases} \frac{1}{2}a_{ij}, & \text{if } i \neq j, \\ -\frac{1}{2}\rho_i, & \text{if } i = j. \ \| \end{cases}$$

5. Cartesian Products of Graphs

In this section we consider isometric embeddings of graphs into metric spaces formed from the Cartesian product of graphs. To begin with, suppose that $G = (V, E)$ is a given connected graph. Define a relation θ on E as follows:

if $e = \{v, w\} \in E$ and $e' = \{v', w'\} \in E$, then $e \, \theta \, e'$ if and only if

$$d_G(v, v') + d_G(w, w') \neq d_G(v, w') + d_G(v', w).$$

This relation was first introduced in an alternative form by Djoković **[15]**. It is easily seen to be well defined, reflexive and symmetric. Let $\hat{\theta}$ denote the transitive closure of θ, and let E_i $(1 \leq i \leq r)$ be the equivalence classes of $\hat{\theta}$.

For each i $(1 \leq i \leq r)$, let G_i denote the graph $(V, E \backslash E_i)$, and let $C_i(1)$, $C_i(2), \ldots, C_i(m_i)$ denote the connected components of G_i. Form the graphs $G_i^* = (V_i^*, E_i^*)$ $(1 \leq i \leq r)$ by letting $V_i^* = \{C_i(1), \ldots, C_i(m_i)\}$ and taking $\{C_i(j), C_i(j')\}$ to be an edge of G_i^* if and only if some edge in E_i joins a vertex in $C_i(j)$ to a vertex in $C_i(j')$. For $v \in C_i(j)$, denote by α_i: $V \to V_i^*$ the natural contraction sending $v \in C_i(j)$ into V_i^*. Define an embedding α: $G \to \prod_{i=1}^{r} G_i^*$, called the **canonical embedding** of G, by

$$\alpha(v) = (\alpha_1(v), \alpha_2(v), \ldots, \alpha_r(v)).$$

In Fig. 3 we illustrate these concepts for a particular graph G.

We call r, the number of factors G_i^* in the canonical embedding of G, the **isometric dimension** of G (for reasons soon to be made clear) and denote it by $\dim_I(G)$.

An isometric embedding $G \xrightarrow{I} \prod_{i=1}^{m} H_i$ is said to be **irredundant** if $|H_i| > 1$

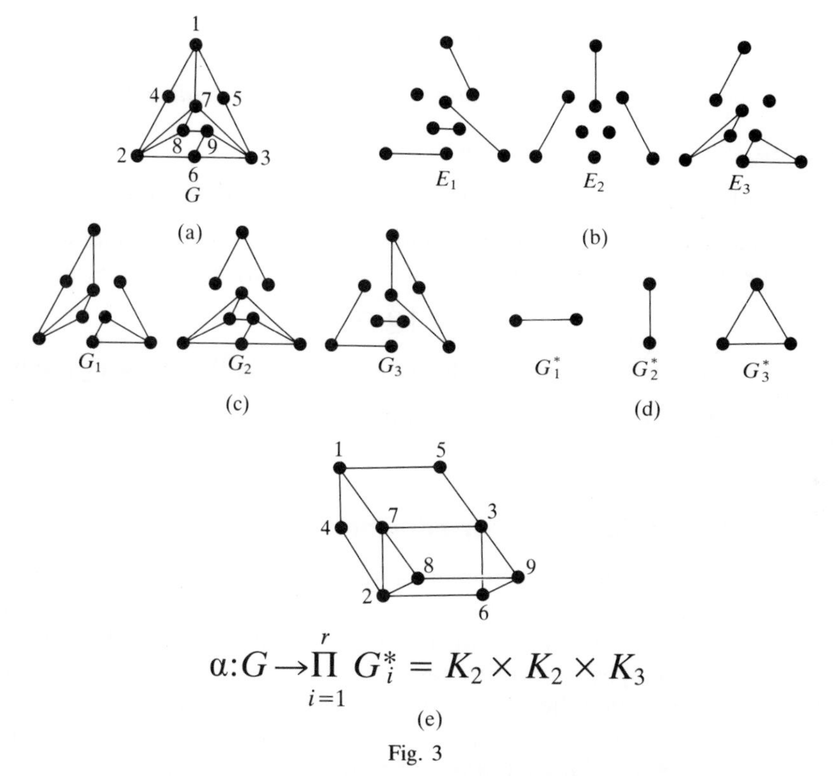

Fig. 3

for all i $(1 \leq i \leq m)$, and for all $h \in H_i$, h occurs as a coordinate value of the image of some $g \in G$. It is not difficult to see that any β: $G \xrightarrow{I} \prod_{i=1}^{m} H_i$ can be made irredundant by discarding unused vertices and factors.

Finally, let us call G **irreducible** if $G \xrightarrow{I} \prod_{i=1}^{m} H_i$ always implies that $G \xrightarrow{I} H_i$ for some i. The following result of **[24]** summarizes the main properties of the preceding concepts:

Theorem 5.1. *If* α: $G \xrightarrow{I} \prod_{i=1}^{r} G_i^*$ *is the canonical embedding, then*

(*i*) α *is isometric*;

(*ii*) α *is irredundant*;

(*iii*) *each factor* G_i^* *is irreducible*;

(*iv*) α *has the largest possible number* $\dim_I(G)$ *of factors among all irredundant isometric embeddings of* G;

(*v*) *the* only *irredundant isometric embedding of* G *into a product of*

$\dim_I(G)$ *factors is the canonical embedding*;

(*vi*) *each factor* H_j *of an irredundant isometric embedding* $G \xrightarrow{I} \prod_{j=1}^{m} H_j$ *embeds canonically into a product of* G_i^**s*. ||

A key fact on which the proof in **[24]** of Theorem 5.1 rests is the following:

Lemma 5.2. *For* $v, w \in V$, *let* P *be a minimal path connecting* v *and* w, *and let* Q *be any path connecting* v *and* w. *Then for any* E_i $(1 \leq i \leq r)$,

$$|P \cap E_i| \leq |Q \cap E_i|. \quad ||$$

As an immediate corollary of Theorem 5.1 we have the following result:

Corollary 5.3. *G is irreducible if and only if G has a single* $\hat{\theta}$-*equivalence class*. ||

6. Almost All Graphs are Irreducible

In the usual random graph model, a random graph $G = (V, E)$ has $V = \{1, 2, \ldots, n\}$ and each pair $\{i, j\}$ is chosen to be an edge with (independent) probability $\frac{1}{2}$. With this model, it is not difficult to show that almost all graphs with n vertices are irreducible as $n \to \infty$. What this means precisely is that the number of graphs with n vertices which are not irreducible, divided by the total number of graphs of this order, tends to zero as n increases. An easy way to see this is to consider the graph G shown in Fig. 4. The dotted lines indicate that an edge may or may not be present. In any case, it is immediate to verify that:

$$d_G(\alpha, x) + d_G(\beta, y) = 2 < 4 = d_G(\alpha, y) + d_G(\beta, x),$$

and $$d_G(\alpha', x) + d_G(\beta', y) = 2 < 4 = d_G(\alpha', y) + d_G(\beta', x),$$

so that if $\{\alpha, \beta\} = e \in E$, $\{\alpha', \beta'\} = e' \in E$, then $e\ \theta\ \{x, y\}$, $e'\,\theta\{x, y\}$, and consequently, $e\ \hat{\theta}\ e'$.

Lemma 6.1. *In almost all random graphs H, for any two pairs of vertices* $\{\alpha, \beta\}$, $\{\alpha', \beta'\}$ *of H there exist vertices x, y of H such that the vertices x and y are connected to* α, β, α', β' *and each other, as shown in Fig. 4.*

Proof. For two fixed (arbitrary) disjoint pairs of vertices $\{\alpha, \beta\}$, $\{\alpha', \beta'\}$ of a random graph H with n vertices, we call H *bad* if no such vertices x and y exist. Since the edges of H are chosen independently and uniformly, the probability that a *given* pair $\{x, y\}$ fails to induce G is at most $1 - 2^{-9}$. Since we can actually form at least $\lfloor \frac{1}{2}(n - 4) \rfloor \geq \frac{1}{2}n - 3$ disjoint candidate pairs $\{x_i, y_i\}$, the probability that all of them fail to induce G is at most

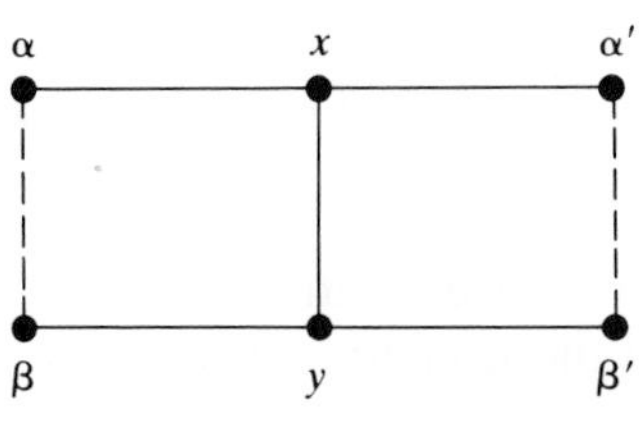

Fig. 4

$(1 - 2^{-9})^{\frac{1}{2}n-3}$. Finally, since there are at most n^4 choices for the initial pairs $\{\alpha, \beta\}$, $\{\alpha', \beta'\}$, the probability that H is bad for *some* choice $\{\alpha, \beta\}$, $\{\alpha', \beta'\}$ is at most $n^4(1 - 2^{-9})^{\frac{1}{2}n-3}$ which certainly tends to 0 as $n \to \infty$. ||

Thus, by Lemma 6.1, in almost all random graphs H any two edges e and e' satisfy $e \ \hat{\theta}\ e'$; that is, H is irreducible. In fact, the preceding argument can be strengthened to show that almost all graphs with n vertices and $cn^{2-\delta}$ edges are irreducible, for an appropriate $\delta > 0$.

It is not difficult to check that, if e and e' are edges belonging to different *blocks* of a connected graph G, then we can never have $e \ \hat{\theta}\ e'$ This implies the following:

Lemma 6.2. *If G is irreducible, then G is 2-connected.* ||

Similar considerations show that if any two edges e and e' of G can be connected by a sequence of triangles in G (that is, there exist triples

$$\{a_1, a_2, a_3\}, \{a_2, a_3, a_4\}, \ldots, \{a_{m-2}, a_{m-1}, a_m\}$$

such that all $\{a_i, a_{i+1}\}$ are edges of G and $e \subseteq \{a_1, a_2, a_3\}$, $e' \subseteq \{a_{m-2}, a_{m-1}, a_m\}$), then G is irreducible. Thus, any triangulated planar graph is irreducible. An analogous result holds for bipartite graphs in which any two edges are connected by a sequence of C_4s.

Irreducible graphs possess many other strong properties which we will not pursue here.

7. Isometric Embeddings into Cubes

In this section we apply some of the preceding theory and investigate graphs G which embed isometrically into Q_r, the r-dimensional cube. Indeed, the early fundamental paper of Djoković **[15]** was devoted to an investigation of such graphs.

Suppose $G \xrightarrow{I} Q_r$ for some r. Then certainly G must be bipartite. Furthermore, suppose $\{v, w\}$ is an edge of G, and assume (without loss of generality) that $v \to (0, 0, \ldots, 0)$ and $w \to (1, 0, \ldots, 0)$. Then any

u with $d_G(u, v) < d_G(u, w)$ must be mapped by the embedding into $(0, u_2, \ldots, u_r)$. In fact, it is not difficult to see that, if

$$d_G(u, v) < d_G(u, w) \text{ and } d_G(v, v) < d_G(v, w),$$

then all points z on any shortest path between u and v must be mapped $z \to (0, z_2, \ldots, z_r)$ by the isometry—that is, z must satisfy $d_G(z, v) < d_G(z, w)$.

The following result of Djoković [15], which can be obtained as a corollary of the previous results on embeddings into Cartesian products, shows that these two necessary conditions are in fact sufficient:

Theorem 7.1. *$G \xrightarrow{I} Q_r$ for some r if and only if*

(*i*) *G is bipartite*;

(*ii*) *for any edge $\{v, w\}$ of G, the set of vertices of G which are closer to v than w is closed under taking shortest paths.* ||

It turns out that when $G \xrightarrow{I} Q_r$, then $\theta = \hat{\theta}$; in fact, this can be used as an alternative characterization, as shown in [24]:

Theorem 7.2. *$G \xrightarrow{I} Q_r$ for some r if and only if*

(*i*) *G is bipartite*;

(*ii*) *the relation θ is transitive.* ||

Another characterization of these graphs was found by Roth and Winkler [30]:

Theorem 7.3. *$G \xrightarrow{I} Q_r$ for some r if and only if*

(*i*) *G is bipartite*;

(*ii*) $n_+(G) = 1$. ||

The proof involves producing a list of forbidden metric subspaces for graphs not isometrically embeddable in Q_r, showing that each of these spaces has at least two positive eigenvalues, and then applying an eigenvalue interlacing theorem.

As noted by Djoković in [15], when $G \xrightarrow{I} Q_r$ then $\dim_I(G)$ is equal to the number of θ-equivalence classes. However, there is also an explicit expression for the value of $\dim_I(G)$, depending only on the signs of the eigenvalues of $\mathbf{D}(G)$ (see [23]):

Theorem 7.4. *If $G \xrightarrow{I} Q_r$, then*

$$\dim_I(G) = n_-(G),$$

the number of negative eigenvalues of the distance matrix $\mathbf{D}(G)$ *of* G.

Proof. First, recall from Theorem 3.3 that $r(G) \geq n_-(G)$. Next, we claim that

$$G \xrightarrow{I} Q_r \Rightarrow n_+(G) = 1.$$

This can be seen by observing (as was done in [**11**]) that no *s are used in the proof of Theorem 3.3, $X_k \cup Y_k$ is always a partition of $V(G)$, and consequently the quadratic form $\sum d_{ij}x_ix_j$ is expressible as a sum of *one* positive square and some negative squares. This implies that $n_+(G) \leq 1$, and since $n_+(G) > 0$ (the trace of $\mathbf{D}(G)$ is zero), we have $n_+(G) = 1$.

We now claim that rank $\mathbf{D}(G) = n + 1$. To see this, observe that, on the one hand,

$$\text{rank } \mathbf{D}(G) = n_-(G) + n_+(G) = n_-(G) + 1 \leq r(G) + 1 \leq r + 1.$$

On the other hand, since G is connected there must exist vertices $v_0, v_1, \ldots, v_r \in V(G)$ such that, if λ: $G \xrightarrow{I} Q_r$ is an isometry, then the set $\{\lambda(v_0), \lambda(v_1), \ldots, \lambda(v_r)\}$ is full-dimensional in Q_r. Thus the submatrix

$$d_G(v_i, v_j) = (d_H(\lambda(v_i), \lambda(v_j))$$

is non-singular, and so rank $\mathbf{D}(G) \geq r + 1$. Consequently, rank $\mathbf{D}(G) = r + 1$, which proves the claim, and

$$n_-(G) = r = r(G) = \dim_I(G),$$

which proves the theorem. ||

It follows from these considerations, for example, that if $G \xrightarrow{I} Q_r$, then

$$\det \mathbf{D}(G) \neq 0 \text{ if and only if } G \text{ is a tree.}$$

8. General Metric Spaces

The problem of embedding graphs isometrically into other graphs is a special case of the more general topic of embedding (finite) metric spaces isometrically into other metric (or semi-metric) spaces. This subject has an extensive literature, some of which can be found in [**1**]–[**10**], [**12**]–[**14**], [**26**]–[**28**] and [**34**]. Many of these more general results apply directly to our problems. For example, it follows from these considerations that if $G \xrightarrow{I} K_3^r$, then $n_+(G) = 1$. The reason for this is as follows.

Let us say that an $r \times r$ distance matrix $\mathbf{D} = (d_{ij})$ is of **negative type** if

$$x_1 + \ldots + x_r = 0 \quad (x_k \in \mathbf{R}) \Rightarrow \sum_{i,j} d_{ij}x_ix_j \leq 0.$$

Similarly, we call **D hypermetric** if

$$x_1 + \ldots + x_r = 1 \quad (x_k \in \mathbf{Z}) \Rightarrow \sum_{i,j} d_{ij} x_i x_j \le 0.$$

Although these two conditions are similar, the latter is actually much stronger (see [**28**]). Not only does it imply the former, but it also implies that the space satisfies the triangle inequality (and many stronger related inequalities), something that a matrix of negative type does not have to. It is not difficult to show that if $\mathbf{D}$ is of negative type, then $n_+(\mathbf{D}) = 1$.

An even more restrictive condition is the following. The matrix **D** is said to be **l_1-embeddable** if **D** can be realized as the distance matrix of a set $X \subseteq \mathbf{R}^m$ for some m, where the distance in $\mathbf{R}^m$ is the usual l_1-metric—that is,

$$d((x_1, \ldots, x_m), (y_1, \ldots, y_m)) = \sum_{k=1}^{m} |x_k - y_k|.$$

It is known (see [**28**]) that any l_1-embeddable space is hypermetric. Of course, if $G \xrightarrow{I} K_2^m$, then G is l_1-embeddable.

It turns out that the properties of l_1-embeddability, hypermetricity and negative type are preserved under taking products, factors and isometric subsets. Thus, for example, since K_3 is of negative type (actually, the matrix

$$\mathbf{D}(K_3) = \begin{pmatrix} 0 & 1 & 1 \\ 1 & 0 & 1 \\ 1 & 1 & 0 \end{pmatrix}$$

is of negative type, which is easy to check), then

$$K_3^m \text{ is of negative type} \Rightarrow G \xrightarrow{I} K_3^m \text{ is of negative type} \Rightarrow n_+(G) = 1,$$

as claimed previously.

An interesting observation due to H. J. Landau (personal communication) is the following. Suppose that X is a metric space with distance matrix **D**. Let $\mathbf{D}^{(k)}$ denote the distance matrix corresponding to the product space X^k. As we have just remarked, if X is of negative type then so is X^k and, consequently, $n_+(\mathbf{D}^k) = 1$ for any k. It turns out rather unexpectedly that the converse holds. In fact, it can be shown that if $n_+(\mathbf{D}^{(2)}) = 1$, then this already implies that X must be of negative type. More generally, if X and Y are finite metric spaces, each having more than one point, and if $n_+(\mathbf{D}(X \times Y)) = 1$, then X and Y must both be of negative type.

9. Concluding Remarks

We show in Fig. 5 a map of some of the metric spaces which have been mentioned in the preceding sections. We conclude by describing a variety of results and open problems which deal with various regions of this map.

To begin with, it was suspected at one time (see [**34**]) that hypermetricity might imply l_1-embeddability. However, this was shown not to be the case, both by Assouad [**1**] and by Avis [**9**], who proved that the

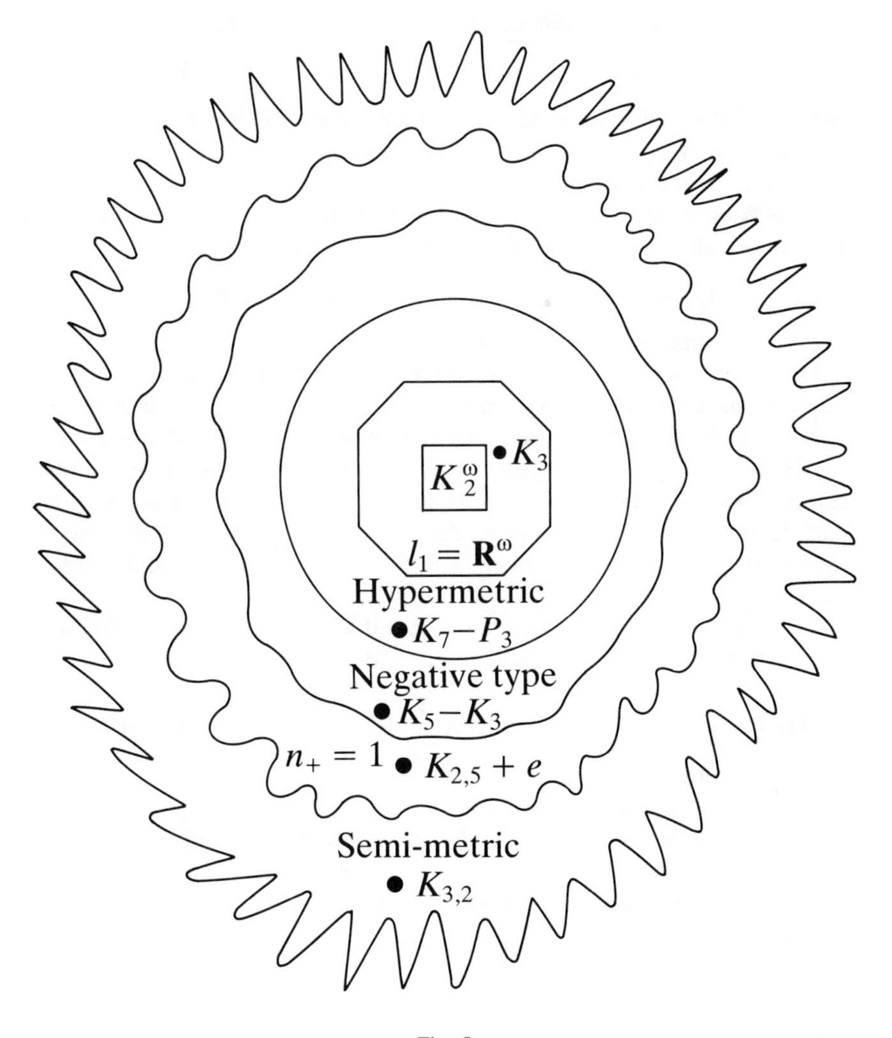

Fig. 5

graph $K_7 - P_3$ is hypermetric but not l_1-embeddable. In the same spirit, it is easy to show that the graph $K_5 - K_3$ is of negative type but not hypermetric (see **[7]**), and that $K_{2,5} + e$ is an example of a graph *not* of negative type which has $n_+ = 1$ (where the added edge e connects the two vertices in the smaller vertex class of $K_{2,5}$; see **[36]**).

We point out that the rather simple graph $K_{3,2}$ is exceptional in several respects. Since $n_+(K_{3,2}) = 2$, $K_{3,2}$ is not of negative type, and is therefore not isometrically embeddable into the n-cube Q_n, or even into $\mathbf{R}^n$. In fact, $K_{3,2}$ is not even a *subgraph* of Q_n. It also turns out that

$$r(K_{3,2}) = 4 > \max(n_+(K_{3,2}), n_-(K_{3,2})) = 3,$$

showing that equality need not hold in Theorem 3.3. It is not known how $r(G)$ behaves for random graphs, but it is natural to guess that $r(G) = |G| - 1$ for almost all large graphs G.

It has been shown by L. Babai and C. Godsil (personal communication) that almost all large graphs G have

$$n_+(G) = (\tfrac{1}{2} + o(1))|G|, \text{ and } n_-(G) = (\tfrac{1}{2} + o(1))|G|,$$

so that presumably Theorem 3.3 almost never holds.

At present no necessary and sufficient conditions are known for a graph to be l_1-embeddable, hypermetric, of negative type or have one positive eigenvalue. An old result of Schoenberg **[31]** shows that a distance matrix $\mathbf{D} = (d_{ij})$ is of negative type if and only if the corresponding distance matrix $\sqrt{\mathbf{D}} = (\sqrt{d_{ij}})$ can be realized by a point set in some Euclidean space. Very recently, Assouad **[3]** has characterized hypermetric spaces in terms of 'deep holes' in certain lattices in Euclidean space.

An interesting problem which has received some attention in the literature is that of determining the quantity $f(n)$, defined by

$$f(n) = \min\{m: |X| = n \text{ and } X \xrightarrow{I} \mathbf{R}^r \text{ for some } r \Rightarrow X \xrightarrow{I} \mathbf{R}^m\},$$

where the l_1-metric is used in $\mathbf{R}^r$ and $\mathbf{R}^m$.

The value of the corresponding function in Euclidean space is clearly $n - 1$, since any set of n points can be embedded isometrically in Euclidean $(n - 1)$-dimensional space. Life is not so simple for the l_1-metric, however. It has been shown by Witsenhausen (personal communication; also see **[8]**) that

$$f(3) = f(4) = 2, f(5) = 3, f(6) \geq 6, f(7) \geq 10, n - 2 \leq f(n) \leq \tbinom{n}{2}.$$

Even the correct order of growth of $f(n)$ is rather mysterious at this point.

It would be highly desirable to have analogues of Djoković's theorem for characterizing those $G \xrightarrow{I} H^r$, for general H. For example, Winkler (personal communication) has shown that $G \xrightarrow{I} K_3^r$ for some r if and only if θ is transitive. At present, however, almost nothing is known in this direction. It would seem fruitful to study the characteristic polynomials of the associated distance matrices of various spaces, rather than just the signs of the eigenvalues. This was initiated for trees in **[12]**, **[16]** and **[21]**. It seems quite likely that our understanding of this whole general area would increase substantially if the corresponding results were known for graphs more general than trees. Good candidates for this would appear to be graphs $G \xrightarrow{I} Q_r$.

Note added in proof. Very recently, Terwilliger and Deza **[32]** have characterized certain classes of hypermetric graphs.

References

1. P. Assouad, Un espace hypermétrique non plongeable dans un espace L^1, *C. R. Acad. Sci. (Paris) (A)* **285** (1977), 361–363; *MR***56**#6327.
2. P. Assouad, Plongements isometriques dans L^1: aspect analytique, *Séminaire d'Initiation à l'Analyse 1979–1980* (Paris 6), Exposé 14.
3. P. Assouad, Sur les inégalités valides dans L^1, *Europ. J. Combinatorics* **5** (1984), 99–112; *MR***86e**:52017.
4. P. Assouad and C. Delorme, Graphes plongeables dans L^1, *C. R. Acad. Sci. (Paris) (A)* **291** (1980), 369–372; *MR***82h**:05043.
5. P. Assouad and C. Delorme, Distances sur les graphes et plongements dans L^1, I, II, to appear.
6. P. Assouad and M. Deza, Espaces métriques plongeables dans un hypercube: aspects combinatoires, *Combinatorics 79* (ed. M. Deza and I. G. Rosenberg), *Ann. Discrete Math.* **8**, North-Holland, Amsterdam, 1980, pp. 197–210; *MR***82h**:05042.
7. P. Assouad and M. Deza, Metric subspaces of L^1, to appear.
8. D. Avis, Hamming metrics and facets of the Hamming cone, Tech. Report SOCS-78.4, School of Computer Science, McGill University, 1978.
9. D. Avis, Extremal metrics induced by graphs, *Combinatorics 79* (ed. M. Deza and I. G. Rosenberg), *Ann. Discrete Math.* **8**, North-Holland, Amsterdam, 1980, pp. 217–220; *MR***81m**:05082.
10. D. Avis, Hypermetric spaces and the Hamming cone, *Canad. J. Math.* **33** (1981), 795–802; *MR***86d**:52007.
11. L. H. Brandenburg, B. Gopinath and R. P. Kurshan, On the addressing problem of loop switching, *Bell System Tech. J.* **51** (1972), 1445–1469.
12. K. L. Collins, Distance matrices of trees, Ph.D. Thesis, M.I.T., 1986.
13. A. K. Dewdney, The embedding dimension of a graph, *Ars Combinatoria* **9** (1980), 77–90; *MR***81m**:05064.
14. M. Deza, Realizability of matrices of distances in unitary cubes (Russian), *Problemy Kibernetiki* **7** (1962), 31–42.
15. D. Z. Djoković, Distance-preserving subgraphs of hypercubes, *J. Combinatorial Theory (B)* **14** (1973), 263–267; *MR***47**#3220.

16. M. Edelberg, M. R. Garey and R. L. Graham, On the distance matrix of a tree, *Discrete Math.* **14** (1976), 23–39; *MR***54**#120.
17. V. Firsov, On isometric embedding of a graph into a Boolean cube (Russian), *Kibernetika (Kiev)* **1** (1965), 95–96; *MR***35**#1500.
18. F. R. Gantmacher, *The Theory of Matrices, Vol. I*, Chelsea, New York, 1959; *MR***21**#6372c.
19. R. L. Graham, On isometric embeddings of graphs, *Progress in Graph Theory* (ed. J. A. Bondy and U.S.R. Murty), *Proc. Waterloo Univ. Silver Jubilee*, Academic Press, Canada, 1984, pp. 307–322; *MR***86f**:05055a.
20. R. L. Graham, A. J. Hoffman and H. Hosoya, On the distance matrix of a directed graph, *J. Graph Theory* **1** (1977), 85–88; *MR***58**#21771.
21. R. L. Graham and L. Lovász, Distance matrix polynomials of trees, *Advances in Math.* **29** (1978), 60–88; *MR***58**#318.
22. R. L. Graham and H. O. Pollak, On the addressing problem for loop switching, *Bell System Tech. J.* **50** (1971), 2495–2519; *MR***44**#6405.
23. R. L. Graham and H. O. Pollak, On embedding graphs in squashed cubes, *Graph Theory and Applications* (ed. Y. Alavi *et al.*), Lecture Notes in Math. **303**, Springer, New York, 1972, pp. 99–110; *MR***48**#8257.
24. R. L. Graham and P. M. Winkler, On isometric embeddings of graphs, *Trans. Amer. Math. Soc.* **288** (1985), 527–536; *MR***86f**:05055b.
25. I. N. Herstein, *Topics in Algebra*, Xerox, Lexington, Mass., 1964; *MR***30**#2028.
26. J. B. Kelly, Metric inequalities and symmetric differences, *Inequalities II* (ed. O. Shisha), Academic Press, New York, 1970, pp. 193–212; *MR***41**#9192.
27. J. B. Kelly, Hypermetric spaces and metric transforms, *Inequalities II* (ed. O. Shisha), Academic Press, New York, 1970, pp. 149–159.
28. J. B. Kelly, Hypermetric spaces, *The Geometry of Metric and Linear Spaces* (ed. L. M. Kelly), Lecture Notes in Math. **490**, Springer, New York, 1975, pp. 17–31; *MR***53**#9161.
29. J. R. Pierce, Network for block switching of data, *Bell System Tech. J.* **51** (1972), 1133–1145.
30. R. L. Roth and P. M. Winkler, Collapse of the metric hierarchy for bipartite graphs, *Europ. J. Combinatorics* **7** (1986), 371–375.
31. I. J. Schoenberg, Metric spaces and completely monotone functions, *Ann. of Math.* **39** (1938), 811–841.
32. P. Terwilliger and M. Deza, The classification of finite connected hypermetric spaces, *Graphs and Combinatorics* **3** (1987), 293–298.
33. H. Tverberg, On the decomposition of K_n into complete bipartite graphs, *J. Graph Theory* **6** (1982), 493–494; *MR***83m**:05111.
34. M. E. Tylkin (M. Deza), On hamming geometry of unitary cubes (Russian), *Doklady Akad. Nauk. SSSR* **134** (1960), 1037–1040; transl. in *Soviet Phys. Dokl.* **5** (1961), 940–943; *MR***22**#13359.
35. P. M. Winkler, Proof of the squashed cube conjecture, *Combinatorica* **3** (1983), 135–139; *MR***85b**:05079.
36. P. M. Winkler, On graphs which are metric spaces of negative type, *Graph Theory and its Applications to Algorithms and Computer Science* (ed. Y. Alavi *et al.*), John Wiley, New York, 1985, pp. 801–810.
37. A. C.-C. Yao, On the loop switching addressing problem, *SIAM J. Comput.* **7** (1978), 515–523; *MR***80c**:68051.

7 Labelings of Graphs

F. R. K. CHUNG

1. Introduction

In this chapter we discuss several interrelated graph labeling problems. These labeling problems have been studied in the past in various different formulations. Typically, the problems can be described as follows: *for a given graph, find the optimal way of labeling the vertices with distinct integers, k-tuples of integers, or group elements subject to certain objectives.* These problems often come up in connection with applications in network addressing, circuit layout or code design.

For example, suppose that we consider labeling the vertices of a graph G by distinct integers. (This can be viewed as arranging the vertices into a line or a linear array.) If we want to find the labeling which minimizes the maximum 'stretch' $b(G)$ over all the edges, we have the so-called *bandwidth problem*. If we want to find the labeling which minimizes the total 'length' sum $s(G)$ of the edges, we have the *minimum-sum problem*. If we want to find the labeling which minimizes the maximum 'overlap' $c(G)$, we have the *cutwidth problem*. (In Fig. 1 we illustrate a tree together with several optimal labelings.) These problems will be rigorously defined in Section 2.

These problems can be considered in the following general framework: label the vertices of a graph G by distinct vertices of a 'host graph' H, and

GRAPH THEORY, 3
ISBN 0-12-086203-4

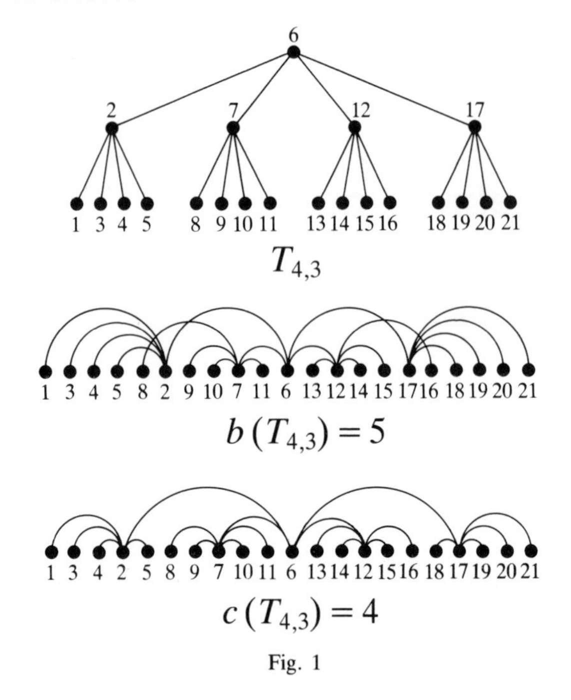

Fig. 1

embed the edges of G into paths of H, subject to any one of the following conditions:

(*i*) the maximum distance in H between adjacent vertices in G is minimized;

(*ii*) the total sum of distances in H between adjacent vertices in G is minimized;

(*iii*) the 'frequency' of edges in H appearing in paths joining adjacent vertices in G is minimized.

For the case of labeling vertices by distinct integers, the host graph is taken to be a path. In the case of labeling vertices by pairs of integers, the host graph is just the grid graph in the plane. When vertices are labeled by binary k-tuples, the labeling provides an embedding into a k-cube as the host graph. Of course, most of the known results in the literature take a path as the host graph.

We shall survey results on various optimal graph labeling problems, and study the properties of these optimal labelings, the relationship with other graph-theoretic parameters, and the algorithmic complexity in determining these optimal labelings for graphs or some special classes of graphs.

2. Definitions and Examples

Let G be a graph with vertex-set $V(G)$ and edge-set $E(G)$. A **labeling** π of G in the host graph H is a one-to-one mapping π from $V(G)$ to $V(H)$. Such a labeling can also be viewed as a placement of the vertices of G into vertices of the fixed host graph H. The following list gives some labelings of interest:

(*i*) The **bandwidth** $b_\pi(G)$ of a labeling π is defined by

$$b_\pi(G) = \max\{d_H(\pi(v), \pi(w)): vw \in E(G)\},$$

where $d_H(v', w')$ denotes the distance between v' and w' in H.

The **bandwidth** $b(G)$ of G is the minimum of $b_\pi(G)$ over all labelings π; that is,

$$b(G) = \min\{b_\pi(G): \pi \text{ is a labeling of } G \text{ in } H\}.$$

(*ii*) A graph G' is said to be a **refinement** of G if G' is obtained from G by a finite number of edge-subdivisions. (For example, in Fig. 2 T' is a refinement of T.) The **topological bandwidth** $b^*(G)$ of G is defined by

$$b^*(G) = \min\{b(G'): G' \text{ is a refinement of } G\}.$$

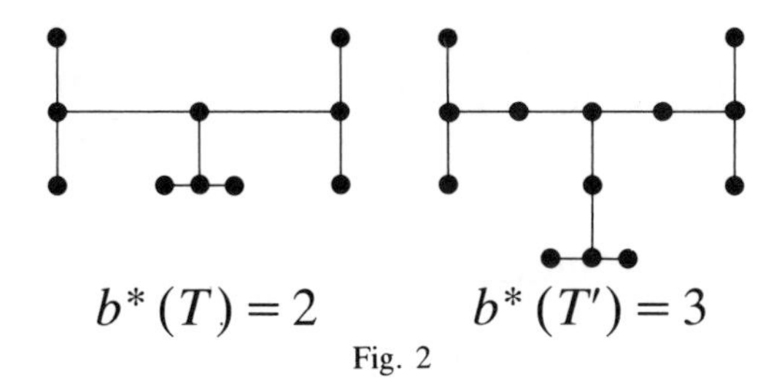

Fig. 2

(*iii*) The **min-sum** $s(G)$ of a labeling π is

$$s(G) = \min\left\{\sum_{vw \in E(G)} d_H(\pi(v), \pi(w)): \pi \text{ is a labeling of } G \text{ in } H\right\}.$$

(*iv*) Suppose we assign a set P_π of paths p in H joining $\pi(v)$ to $\pi(w)$ for all $vw \in E(G)$—that is, $P_\pi = \{p(\pi(v), \pi(w)): vw \in E(G)\}$. We define the **cutwidth** $c_\pi(G)$ of a labeling π to be the maximum number of appearances of an edge e of H in the paths in P_π. In other words,

$$c_\pi(G) = \max_{e \in E(H)} |\{vw \in E(G): e \in E(p(\pi(v), \pi(w))), p(\pi(v), \pi(w)) \in P_\pi\}|.$$

The **cutwidth** $c(G)$ of G is then defined to be

$$c(G) = \min\{c_\pi(G): \pi \text{ is a labeling of } G \text{ in } H\}.$$

When the host graph H is a path with vertices $\{1, 2, \ldots\}$, the labeling of G is a mapping from $V(G)$ into the set of positive integers. Such a labeling is often called a **numbering** of G. The distance between two numbers is just the absolute value of the difference. The bandwidth of a numbering π can therefore be written as

$$b_\pi(G) = \max\{|\pi(v) - \pi(w)|: vw \in E(G)\}.$$

Also the cutwidth of a numbering π can be written as

$$c_\pi(G) = \max_i |\{vw \in E(G): \pi(v) \leqslant i < \pi(w)\}|.$$

We conclude this section with a table showing various optimal numberings for paths P_p, circuits C_p, stars $K_{1,p}$, and complete graphs K_p (see **[13]**):

	b	b^*	s	c
P_p	1	1	$p - 1$	1
C_p	2	2	$2(p - 1)$	2
$K_{1,p}$	$\lfloor \frac{1}{2}p \rfloor$	$\lfloor \frac{1}{2}p \rfloor$	$\lfloor \frac{1}{4}p^2 \rfloor$	$\lfloor \frac{1}{2}p \rfloor$
K_p	$p - 1$	$p - 1$	$p(p^2 - 1)$	$\lfloor \frac{1}{4}p^2 \rfloor$

3. The Bandwidth Numbering

Among all graph labeling problems, bandwidth numberings of graphs have attracted the most attention in the literature. The bandwidth problem originated in the 1950s in the form of finding a matrix equivalent to a given matrix so that all the non-zero entries lie within a narrow band about the main diagonal. Harper **[29]** investigated the bandwidth numberings for n-cubes; these are related to the design of error-correcting codes subject to minimizing the maximum absolute error. Since then there have been a large number of papers on this subject, references for most of which can be found in a recent survey **[5]** on bandwidth numberings.

One of the interesting problems on bandwidth is to characterize graphs with large or small bandwidth. What does it take to force a large bandwidth in a graph? There are two known factors which increase the bandwidth of a graph — density (roughly, the ratio of the total number of vertices to the maximum distance between pairs of vertices), and subtrees which are refinements of complete binary trees. As we shall see, either of these two conditions is sufficient but not necessary for forcing up the bandwidth of a graph. We list some known results as well as some obvious (but unpublished) facts on lower bounds for bandwidth numberings.

The following two density lower bounds can be derived in a straightforward way. The first appeared in the early paper of Harper **[29]**, whereas the second one was a folklore theorem discovered by many people independently (see **[11]**):

Theorem 3.1. *Let ∂S denote the set of vertices in S adjacent to some vertices in $V(G) - S$. Then*

$$b(G) \geqslant \max_{k} \min_{|S| = k} |\partial S|. \quad \|$$

Theorem 3.2. *Let $D(G)$ denote the diameter of G. Then*

$$b(G) \geqslant (|V(G)| - 1)/D(G). \quad \|$$

If G is a graph with p vertices and diameter D, we define the **density** $ds(G)$ to be $(p - 1)/D$. The **local density** of G is defined to be the maximum density of all subgraphs of G, and can sometimes provide better lower bounds for the bandwidth of G.

Theorem 3.3. $b(G) \geqslant \max\{ds(G')$: *$G'$ is a subgraph of G*$\}$. $\|$

From Theorem 3.3 we can easily deduce the following result, by taking G' to be a star graph (see **[11]**).

Theorem 3.4. *If Δ denotes the maximum degree in G, then*

$$b(G) \geqslant \tfrac{1}{2}\Delta. \quad \|$$

Sysło and Zaks **[43]** proved that the local density lower bound in Theorem 3.3 determines the bandwidth of a special class of trees called *caterpillars*. A **caterpillar** is a tree in which the removal of all end-vertices leaves a path P called the **spine**. The end-edges are called **leaves**.

Theorem 3.5. *Let G be a caterpillar whose spine consists of the vertices $u_1, \ldots, u_m$. If the vertex-degrees $\rho(u_1), \ldots, \rho(u_m)$ form a monotone sequence, then*

$$b(G) = \max_{i}\{\lceil(|V(G_i)| - 1)/D(G_i)\rceil\},$$

where G_i is the subcaterpillar formed by $u_1, \ldots, u_i$ and all vertices adjacent to these u_j, for $i = 1, \ldots, m$. $\|$

In fact, the bandwidths of all caterpillars depend only on the local density:

Theorem 3.6. *If G is a caterpillar, then*

$$b(G) = \max_{G'}\{\lceil(|V(G')| - 1)/D(G')\rceil\},$$

where G' consists of a subpath P' of the spine and all vertices adjacent to vertices in P'.

Proof. Suppose that π is a numbering of G with $b_\pi(G) = b(G) = b$. We can normalize π so that

(*i*) all vertices u_i, $1 \leqslant i \leqslant m$, on the spine are mapped to $bi + 1$;

(*ii*) the leaves adjacent to u_i occupy all the numbers from $\pi(u_i) - r_i$ to $\pi(u_i) + s_i$, except $\pi(u_i)$, with u_i on the spine;

(*iii*) the number of r_i with maximum value is minimized—the maximum value is $b - 1$, except for r_1 which has value b;

(*iv*) some r_i has maximum value.

Now we start from u_j with r_j of maximum value and search for the first k ($>j$) such that $s_k + r_{k+1} < b - 1$ or $s_k = b - 1$. If no such k exists, we take $k = m$. If k satisfies $s_k + r_{k+1} < b - 1$ or $k = m$ and $s_m < b$, then we can modify π by increasing r_t and s_t by 1 for $j \leqslant t \leqslant k$ and thus decrease the number of maximum r_i, contradicting assumption (*iii*). We may assume that $s_k = b - 1$ (or $s_m = b$) and $s_t + r_{t+1} = b - 1$ for $j \leq t \leq k$. Thus, by taking G' with the spine P' consisting of $u_j, \ldots, u_k$, we have

$$(|V(G')| - 1)/D(G') \geqslant \lceil (b(k - j + 1) + 1)/(k - j + 2) \rceil = b,$$

so $b(G) \leqslant \lceil (|V(G')| - 1)/D(G') \rceil$ for some G'. Combining this with Theorem 3.3, we deduce Theorem 3.6. ||

Let $T_{2,k}$ denote the k-level complete binary tree in which the ith level consists of 2^{i-1} vertices and each vertex in level $i < k$ has two 'sons' at level $i + 1$. Then the bandwidth of $T_{2,k}$ depends on k, as follows:

Theorem 3.7. $b(T_{2,k}) = \lceil (2^{k-1} - 1)/(k - 1) \rceil$.

Proof. By Theorem 3.2 we have $b(T_{2,k}) \geqslant x$, where $x = \lceil (2^{k-1} - 1)/(k - 1) \rceil$. It suffices to find a numbering π with $b_\pi(T_{2,k}) = x$. We define π by specifying in increasing order the vertices chosen from the left half (descendants of the left son) of $T_{2,k}$—the first one chosen is labeled 1, and so on; the right half is then chosen in the reverse order of the left half. We first choose x vertices in the left-most part of the bottom (kth) level. Then we choose the y_1 fathers of these x vertices and then choose the $x - y_1$ left-most (unchosen) vertices of the bottom level. We continue to choose x vertices at each step, in such a way that the y_i fathers of previously chosen ones are chosen first, and then the bottom level left-most (unchosen) ones. It is not hard to check that, for $i > 2$, $y_i < \frac{1}{2}x + i$. We have $y_i < x$ if $i \leqslant \frac{1}{2}x$, which is true for $k > 5$. (For $k \leqslant 5$, the bandwidth can be verified directly.) So we can proceed until at most x vertices are left. The remaining vertices are chosen at the final step. ||

The complete k-level t-ary tree $T_{t,k}$ is also determined by its density; the proof is quite similar to that of Theorem 3.7 and will not be included here.

Theorem 3.8. $b(T_{t,k}) = \left\lceil \dfrac{t(t^{k-1} - 1)}{2(k - 1)(t - 1)} \right\rceil$. ||

For trees with bounded degree we have an upper bound which is within a constant factor of the worst-case density lower bounds:

Theorem 3.9. *Suppose that T is a tree on p vertices with bounded degree ρ. Then*

$$b(T) \leqslant 5 \frac{p}{\log_\rho p}.$$

Proof. This upper bound is established by a numbering scheme which uses a result in **[6]**—namely, for a given tree T and any positive value k we can find a vertex v such that there is a subforest F, formed by removing v from T, with $k \leqslant |V(F)| < 2k$. Now we first find a separator set S consisting of $\lceil \log_\rho p - \log_\rho \log_\rho p + 1 \rceil = y$ vertices whose removal separates T into at most y forests $F_1, F_2, \ldots, F_y$, such that $p/y \leqslant |V(F_i)| < 2p/y$, for all i. Furthermore, we may assume that the forests F_i are arranged in increasing order of the number of edges from F_i to T. Now we partition the integers into $3y$ (increasing) blocks B_j, each consisting of $\lceil p/y \rceil$ consecutive numbers. The labeling can be described recursively as follows:

—the numbers in B_1 and B_2 are used to label F_1;

—B_3 is used for labeling the vertices adjacent to vertices in F_1;

—B_4 and B_5 are used to label the (so-far unlabeled) vertices of F_2;

—B_6 is used for labeling the unlabeled vertices adjacent to vertices of F_2 and unlabeled vertices adjacent to vertices labeled by numbers in B_3.

In general, B_{3i+1} and B_{3i+2} are used for labeling the so-far unlabeled vertices of F_{i+1}, and B_{3i+3} is used for labeling the (unlabeled) neighbors of F_{i+1} and the (unlabeled) neighbors of vertices of B_{3i}. Since $\bigcup_{i \leqslant j} F_i$ has at most j vertices in the separating set S, the number of neighbors that B_{3j+3} used for labeling is at most

$$\rho^{j-1} + \rho^{j-2} + \ldots + 1 = \frac{\rho^j - 1}{\rho - 1} \leqslant \rho^y \leqslant \frac{p}{y}.$$

This labeling obviously has bandwidth $4\dfrac{p}{y} \leqslant 5\dfrac{p}{\log_\rho p}$. ||

The density or local density is not an upper bound for graphs in general. There are graphs with large bandwidth but small density. The following result comes from [**12**]:

Theorem 3.10. *For each integer k, there is a tree with local density* 1 *or* 2 *and bandwidth at least k.*

Proof. We consider a refinement T of the complete binary tree $T_{2,2k}$ with $2k$ levels. An edge joining a vertex of the ith level and a vertex of the $(i + 1)$th level is replaced by a path of length 3^{2k-i}. It can be easily checked that T has local density at most 2. The bandwidth of T is not less than $b^*(T_{2,k})$, which is k (see [**12**] and also Section 6). ||

Theorem 3.11. *Any graph which contains a refinement of $T_{2,k}$ must have bandwidth at least $\frac{1}{2}k$.* ||

Of course, there are graphs with large bandwidth which do not contain large complete binary trees. This leads to the following interesting problem:

Problem 3.1. Suppose that the local density of G is at most c_1, and that G does not contain any refinement of the complete binary tree with c_2 levels. Is the bandwidth of G bounded above by a constant depending only on c_1 and c_2?

Note added in proof. This problem was recently answered in the negative by Chung and Seymour.

Chvátalová and Opatrný [**14**] proved a somewhat weaker result:

Theorem 3.12. *Suppose that T is an infinite countable tree such that*

(*i*) *the maximum degree satisfies $\Delta \leqslant c_1$;*

(*ii*) *the number of edge-disjoint semi-infinite paths in T is at most c_2;*

(*iii*) *T does not contain (as a subgraph) a refinement of a complete binary tree of c_3 levels. Then T has a refinement T' with finite bandwidth depending only on c_1, c_2 and c_3.* ||

Another useful observation for dealing with bandwidth involves the graph P_n^k, the kth power of the path P_n, in which two vertices v, w are adjacent if and only if $0 < |v - w| \leqslant k$. It follows immediately that $b(G) \leqslant k$ if and only if $G \subseteq P_n^k$. Many relationships between the bandwidth and other graph invariants were derived using this observation (see [**11**]); for example:

Theorem 3.13. *Suppose that G has p vertices, q edges, connectivity κ, independence number β_0, and degree-sequence $\rho_1 \leqslant \rho_2 \leqslant \ldots \leqslant \rho_p$. Then:*

$$b(G) \geqslant p - \tfrac{1}{2}(1 + ((2p - 1)^2 - 8q)^{\frac{1}{2}});$$
$$b(G) \geqslant \kappa;$$

$$b(G) \geqslant \frac{p}{\beta_0} - 1;$$

$$b(G) \geqslant \max_j \max\{\rho_j - \lfloor \tfrac{1}{2}(j-1) \rfloor, \tfrac{1}{2}\rho_j\}.$$

Proof. Observe that $|V(G_1)| = |V(G_2)|$, and that $G_1 \subseteq G_2$ implies $|E(G_1)| \leqslant |E(G_2)|$, $\kappa(G_1) \leqslant \kappa(G_2)$, $\beta_0(G_1) \geqslant \beta_0(G_2)$ and $\rho_j(G_1) \leqslant \rho_j(G_2)$, for each j. By noting that for $k = b(G)$ we have $G \subseteq P_p^k$ and $|E(P_p^k)| = \frac{1}{2}k(2p - k - 1)$, we get $\kappa(P_p^k) = k$, $\beta_0(P_p^k) = \lceil (p-1)/k \rceil$, $\rho_j(P_p^k) = \min\{p - 1, k + \lfloor \frac{1}{2}(j-1) \rfloor, 2k\}$. ||

We remark that parts of Theorem 3.13 were also proved by A. K. Dewdney **[5]**, **[13]**. For completeness, we include here the bandwidths of a few special graphs (see **[11]**, **[18]** and **[29]**):

Theorem 3.14. (*i*) $b(K_{r,s}) = \frac{1}{2}(r - 1) + s$, *for* $r \geqslant s$;

(*ii*) $b(K_{r_1, r_2, \ldots, r_k}) = p - \lceil \frac{1}{2}(r_1 + 1) \rceil$;

(*iii*) $b(Q_n) = \sum_{k+1}^{n-1} \binom{k}{\lfloor \frac{1}{2}k \rfloor}$, *where* Q_n *is the n-cube.* ||

4. Bandwidth Algorithms

Before we discuss the algorithmic aspects of the bandwidth problem, we shall give a very brief introduction. In general, algorithms are step-by-step procedures for producing solutions for problems. A **polynomial-time algorithm** is an algorithm which always generates a solution in time $p(n)$, for some polynomial function p, where n denotes the input length. A problem is considered to be **intractable** if it is so hard that no polynomial algorithm can possibly solve it. Many problems that are not known to be either provably intractable or provably polynomial turn out to be so-called ***NP*-complete** (non-deterministic polynomial-time complete), which were first introduced in the early 1970s (see **[20]**). A problem in the *NP*-complete class has the property that if a polynomial-time algorithm is ever found for it, then all the problems in this class must also have polynomial-time algorithms. This *NP*-complete class of problems includes many 'classical' problems, such as the traveling salesman problem, the Hamiltonian circuit problem, integer linear programming, and many others. The question of whether or not the *NP*-complete problems are intractable is now considered to be one of the foremost open questions in theoretical computer science.

Papadimitriou **[36]** first proved that the bandwidth problem is *NP*-complete, by showing that it is polynomial transformable from the 3-partition problem. The **3-partition problem** is one of the basic *NP*-complete problems, and can be described as follows:

Suppose that A is a set of $3m$ elements with total weight $\sum_{a \in A} \omega(a) = mB$. Can A be partitioned into m disjoint 3-sets $A_i (i = 1, \ldots, m)$ such that $\sum_{a \in A_i} \omega(a) = B$ for each i?

Theorem 4.1. *The bandwidth problem is NP-complete.* ||

Garey *et al.* [**19**] proved that the bandwidth problem remains *NP*-complete for trees with maximum degree 3. The bandwidth of a graph G is 1 if and only if each connected component of G is an isolated vertex or a path. In [**19**] there is a linear algorithm for testing if $b(G) = 1$ or 2. Saxe [**39**] has shown that, for fixed k, the problem of determining $b(G) \leqslant k$ can be solved in polynomial time.

There are several approximation algorithms for the bandwidth problem, such as the Cuthill–McKee algorithm [**16**] and the Gibbs–Poole–Stockmeyer algorithm [**23**]. One way to judge the approximation algorithms is to investigate the worst-case performance bounds (that is, the bounds for the ratio of the bandwidth generated by the algorithm and the 'optimum' bandwidth of the graph), or the average performance bounds. However, for these approximation algorithms neither the worst-case performance nor the average performance bound has been extensively analyzed.

5. Bandwidth Labeling in Higher Dimensions

The (two-dimensional) **grid graph** is the graph whose vertex-set consists of pairs of integers, and whose edge-set consists of $\{(i, j), (i + 1, j)\}$ and $\{(i, j), (i, j + 1)\}$, for all integers i, j. Graph labeling problems whose grid graph is the host graph often arise in the formulation of circuit layout models involved in VLSI design or optimization (see [**2**], [**3**]). In particular, the bandwidth of the layout is directly related to the performance of the circuit if the length of the edge is proportional to the propagation delay through the wire, which must be smaller than the period of the system clock. In models with higher-dimensional grid graphs as the host graph, bandwidth problems correspond to layout problems in multi-layer circuits.

Before we proceed to discuss bandwidth results in grid graphs, we remark that the distance of two vertices (a, b) and (c, d) in the grid graph is defined to be $|a - c| + |b - d|$. The *k-dimensional grid graph* G_i has vertex-set consisting of all k-tuples of integers, and edge-set consisting of $\{(a_1, \ldots, a_i, \ldots, a_k), (a_1, \ldots, a_i + 1, \ldots, a_k)\}$, for all a_i and i.

Analogous to the bandwidth numbering problem, we have the following density lower bound:

Theorem 5.1. *If G is a graph with p vertices and diameter D, then the bandwidth of G in the 2-dimensional grid graph is bounded below by $(p^{\frac{1}{2}} - 1)/D$.*

Proof. Let f denote the bandwidth labeling from $V(G)$ to the grid graph G_2. Consider the lines

$$\begin{aligned} L_1&: x + y = \min\{a: (a, b) = f(v), \quad v \in V(G)\} = p_1; \\ L_2&: x + y = \max\{a: (a, b) = f(v), \quad v \in V(G)\} = p_2; \\ L_3&: x - y = \min\{b: (a, b) = f(v), \quad v \in V(G)\} = q_1; \\ L_4&: x - y = \max\{b: (a, b) = f(v), \quad v \in V(G)\} = q_2. \end{aligned}$$

Clearly all vertices of G are embedded (in G_2) in the rectangle bounded by these four lines. This rectangle contains at most $(|p_1 - p_2| + 1)(|q_1 - q_2| + 1)$ vertices. Without loss of generality, we may assume that $|p_1 - p_2| \geqslant |q_1 - q_2|$. It is not hard to check that there are two vertices of G embedded in G_2 with distance $|p_1 - p_2|$ in G. Therefore $bD \geqslant |p_1 - p_2|$, where b denotes the bandwidth. Since

$$(|p_1 - p_2| + 1)^2 \geqslant (|p_1 - p_2| + 1)(|q_1 - q_2| + 1) \geqslant p,$$

we have $b \geqslant (p^{\frac{1}{2}} - 1)/D$. ||

We also have the following theorem:

Theorem 5.2. *Let $b_2(G)$ denote the bandwidth of G in the 2-dimensional grid graph. Then*

$$b_2(G) \geqslant \max\left\{\left\lceil \frac{|V(G')|^{\frac{1}{2}} - 1}{D(G')} \right\rceil : G' \text{ is a subgraph of } G\right\}. \ ||$$

The higher-dimensional cases can be proved in a similar way.

Theorem 5.3. *If G is a graph with p vertices and diameter D, then the bandwidth of G in the k-dimensional grid graph is bounded below by $(p^{1/k} - 1)/D$.* ||

Theorem 5.4. *Let $b_k(G)$ denote the bandwidth of G in the k-dimensional grid graph. Then*

$$b_k(G) \geqslant \max\left\{\left\lceil \frac{|V(G')|^{1/k} - 1}{D(G')} \right\rceil : G' \textit{ is a subgraph of } G\right\}. \ ||$$

In the case of embedding complete binary trees in the 2-dimensional grid graph, the density lower bound is essentially $n^{\frac{1}{2}}/\log n$, where n is the input length. Paterson *et al.* **[37]** proved that the bandwidths b_2 of such trees are within a constant factor of the density lower bounds:

Theorem 5.5. *If $n = 2^{k+1} - 1$, then $b_2(T_{2,k}) = \Omega(n^{\frac{1}{2}}/\log n)$; in other words, $b_2(T_{2,k})$ is bounded above and below by expressions of the form $cn^{\frac{1}{2}}/\log n$.* ||

Problem 5.1. Evaluate $b_k(T_{i,j})$, for general k.

Bhatt and Leighton [3] used a general technique to obtain upper bounds for the bandwidth b_2 for binary trees (not necessarily complete) and planar graphs:

Theorem 5.6. *If T is a binary tree (that is, a tree with maximum degree 3), then*

$$b_2(T) \leqslant \Omega(n^{\frac{1}{2}}/\log n). \; \|$$

Theorem 5.7. *If G is a planar graph, then*

$$b_2(G) \leqslant O\left(\frac{n^{\frac{1}{2}} \log n}{\log \log n}\right). \; \|$$

Problem 5.2. Find good bounds for $b_k(G)$, when G is planar or of bounded genus.

The algorithmic problem of finding optimal labelings in grid graphs turns out to be considerably harder than the case of embedding into a line, as the following result by Bhatt and Cosmadakis [2] shows:

Theorem 5.8. *Given a binary tree T, the problem of deciding whether $b_2(T) = 1$ is NP-complete.* ||

Unless otherwise specified, in the remainder of the chapter we discuss only labelings with a path as the host graph.

6. Topological Bandwidth

We recall that the topological bandwidth $b^*(G)$ of a graph G is the minimum bandwidth $b^*(G')$ over all refinements G' of G. It follows from the definition that $b^*(G') \geqslant b^*(G)$ for a refinement G' of G. Strict inequality can occur, as shown by Fig. 2.

The topological bandwidth problem can be related to optimization problems arising in some models of VLSI design in which vertices of degree 2 (interpreted as 'drivers' or 'repeaters') are inserted to help minimize the length of the edges. There is also a sparse-matrix version of the topological bandwidth problem [34]. Let $\mathbf{A}$ be a matrix arising from a linear system $\mathbf{Ax} = \mathbf{b}$. The bandwidth problem is just the problem of finding a permutation matrix $\mathbf{P}$ such that $\mathbf{PAP}^{\mathrm{T}}$ has all of its non-zero entries close to the main diagonal. The topological bandwidth problem can then be viewed as the problem of further narrowing the 'distance' to the main diagonal by repeatedly using the following operations: replace a term $a_{ij}x_j$ by a new variable y and add a new equation $a_{ij}x_j = y$.

There are fewer results on the topological bandwidth than on the bandwidth of graphs. The following results are taken from [**12**], [**35**] and [**42**].

Theorem 6.1. (*i*) $b^*(G) \geqslant \frac{1}{2}\Delta$;

(*ii*) $b^*(G) \geqslant \min_{v \in V(G)} \deg(v)$. ||

Theorem 6.2. (*i*) *For the k-level complete binary tree* $T_{2,k}$, $b^*(T_{2,k}) = \lceil \frac{1}{2}k \rceil$;

(*ii*) $b^*(K_{2,s}) \geqslant \lfloor \frac{1}{2}(s + 3) \rfloor$. ||

Theorem 6.3. (*i*) *There is an* $O(n \log n)$ *algorithm for computing the topological bandwidth of binary trees*;

(*ii*) *the problem of determining the topological bandwidth of a graph is NP-complete.* ||

We conclude this section with the following problem:

Problem 6.1. Determine the computational complexity of the topological bandwidth for trees.

Partial results in this direction can be found in Dewdney [**17**].

7. The Minimum Sum Problem

Instead of minimizing the maximum 'stretch' over all the edges, the minimum sum problem is to search for a labeling which minimizes the total sum of the stretches of the edges. In the case of labeling vertices by integers (or when the host graph is a path), the minimum sum problem is just that of minimizing $s(G) = \sum_{vw \in E(G)} |\pi(v) - \pi(w)|$ over all numberings π. The minimum sum problem was first investigated by Harper [**27**], who determined the optimal labelings for the class of n-cubes which are associated with designing error-correcting codes with minimum average absolute errors:

Theorem 7.1. *For the n-cube* Q_n, $s(Q_n) = 2^{n-1}(2^n - 1)$. ||

Seidvasser [**40**] studied the maximum min-sum $s(T)$ over all trees with p vertices and maximum degree ρ. His results were further improved by Iordanskiĭ [**30**]:

Theorem 7.2. *For any tree T on p vertices and maximum degree ρ, we have*

$$s(T) \leqslant c\,\frac{\rho p \log p}{\log \rho}, \text{ for some constant } c,$$

and there exist trees for which $s(T)$ lies within a constant factor of this upper bound, for all p and ρ. ||

Chung [**6**] answered a question of Cahit [**4**] in determining $s(T_{2,k})$ for the complete k-level binary tree $T_{2,k}$:

Theorem 7.3. $s(T_{2,k}) = 2^k(\frac{1}{3}k + \frac{5}{18}) + (-1)^k\frac{2}{9} - 2$ *for* $k \geqslant 2$. ||

For general t ($t > 2$), it is much more difficult to derive an explicit expression for $T_{t,k}$. A recurrence relation for $s(T_{3,k})$ can be described as follows (see [**6**]):

Theorem 7.4. *Let $k \geqslant 3$, and let $f(k)$ be the integer l which satisfies $(l-1)3^{l-2} + l + 1 \leqslant k < l3^{l-1} + l$. Then*

$$s(T_{3,k}) = 3^{k-2}(2k - \tfrac{1}{2}) - \tfrac{1}{2} + k - f(k) + s(T_{3,k-1}).\ \|$$

As to the algorithmic problems, Garey *et al.* [**22**] have proved the *NP*-completeness of the min-sum problem:

Theorem 7.5. *The problem of determining the minimum sum labeling for a general graph is NP-complete.* ||

In the special case of a tree, Goldberg and Klipker [**24**] have an $O(p^3)$ algorithm for determining a min-sum labeling for a tree on p vertices. Shiloach [**41**] improved the algorithm to one which has $O(p^{2.2})$ in running time. This was further improved by Chung in [**8**]:

Theorem 7.6. *If* $\lambda = \log 3/\log 2 \simeq 1.6$, *there is an $O(p^\lambda)$ algorithm for finding the min-sum labeling for trees on p vertices.* ||

8. The Cutwidth of Graphs

The cutwidth problem deals with the number of edges passing over a vertex, when all vertices are arranged in a line; note that $c(G) = \min_\pi \max_i |\{vw \in E(G): \pi(v) \leqslant i < \pi(w)\}|$. The cutwidth often corresponds to the area of the layout in VLSI design.

Lengauer [**32**] evaluated the cutwidths for the complete k-level t-ary trees $T_{t,k}$:

Theorem 8.1. $c(T_{t,k}) = \lceil \frac{1}{2}(k-1)(t-1) \rceil + 1$, *for* $k \geqslant 3$. ||

The values of cutwidths for trees are at least as large as their topological bandwidths (see **[9]**, **[34]**):

Theorem 8.2. $b^*(G) \leqslant c(G)$. ||

For general graphs, $b^*(G)$ and $c(G)$ can be quite different; for example, $b^*(K_p) = p - 1$ and $c(K_p) = \lfloor \frac{1}{4}p^2 \rfloor$. However, for a tree T it was proved in **[9]** that the values of $b^*(T)$ and $c(T)$ are quite close:

Theorem 8.3. $b^*(T) \leqslant c(T) \leqslant b^*(T) + \log_2 b^*(T) + 2$. ||

These bounds are 'almost' best possible in the sense that, for each n, there exists a tree T with $b^*(T) = n$ and $c(T) \geqslant n + \log_2 n - 1$, and there exists a tree T' with $b^*(T') = c(T') = n$.

Stockmeyer (see **[20]**) proved the *NP*-completeness of the cutwidth problem:

Theorem 8.4. *The cutwidth problem for general graphs is NP-complete.* ||

Recently, Makedon *et al.* **[35]** proved that the cutwidth problem remains *NP*-complete when restricted to graphs with maximum degree 3. Gurari and Sudborough **[26]** proved that, for fixed k, the problem of determining whether the cutwidth is at most k is polynomial:

Theorem 8.5. *There is an $O(n^k)$ algorithm for deciding whether $c(G) \leqslant k$ for a graph G.* ||

Yannakakis **[44]** recently resolved the complexity problem for determining the cutwidth for trees:

Theorem 8.6. *There is an $O(n \log n)$ algorithm for determining the cutwidth for a tree.* ||

There are several results involving the relations of cutwidth with other graph invariants, such as the search numbers and graph pebblings (see **[33]**, **[38]**, **[42]**), which we shall not discuss here.

9. Concluding Remarks

Graph labeling is an active area within graph theory, having connections with a variety of application-oriented areas such as VLSI optimization, data structure and data representation. In the past few years, many new directions and new results have been developed while many questions remain unresolved. Here we mention several general directions for future research.

(*i*) Find good approximation algorithms for the *NP*-complete labeling problems mentioned in Sections 3–8. Such algorithms are very desirable because of the advancement in integrated circuit technology and the need in design automation.

(*ii*) Characterize graphs with good labelings—for example, the problem posed in Section 3 (does bounded density and no large complete binary subtree imply small bandwidth?) is one of the problems in characterizing graphs with small bandwidth.

(*iii*) Until now, most of the work has been concentrated on the case in which the host graph is a path. Results concerning the grid graphs as host graphs are quite recent and are known only for the bandwidth labelings. Most labeling problems using grid graphs as the host graphs have not been studied. Many of these problems are of particular interest in rectilinear network layout designs. There are many other good candidates for host graphs, such as trees and circuits; in taking a circuit as host graph, we deal with integers modulo n as labels. Taking various trees as the host graph is associated with problems in data structures, and has a special appeal of its own.

References

1. D. Adolfson and T. C. Hu, Optimal linear ordering, *SIAM J. Appl. Math.* **25** (1973), 403–423; *MR***49**#10349.
2. S. N. Bhatt and S. Cosmadakis, The complexity of minimizing wire lengths in VLSI layouts, preprint.
3. S. N. Bhatt and F. T. Leighton, A framework for solving VLSI graph layout problems, *J. Comput. System Sci.* **28** (1984), 300–343; *MR***86g**:68139.
4. I. Cahit, A conjectured minimum valuation tree, *SIAM Review* **19** (1977), 164.
5. P. Z. Chinn, J. Chvátalová, A. K. Dewdney and N. E. Gibbs, The bandwidth problem for graphs and matrices—a survey, *J. Graph Theory* **6** (1982), 223–254; *MR***84g**:05100.
6. F. R. K. Chung, A conjectured minimum valuation tree, Problems and solutions, *SIAM Review* **20** (1978), 601–604.
7. F. R. K. Chung, Some problems and results in labelings of graphs, *The Theory and Applications of Graphs* (ed. G. Chartrand *et al.*), John Wiley, New York, 1981, pp. 255–263.
8. F. R. K. Chung, On optimal linear arrangements of trees, *Computers and Math. with Applications* **10** (1984), 43–60; *MR***85b**:05065.
9. F. R. K. Chung, On the cutwidth and the topological bandwidth of a tree, *SIAM J. Discrete Alg. Methods* **6** (1985), 268–277; *MR***86k**:05036.
10. F. R. K. Chung and R. L. Graham, On graphs which contain all small trees, *J. Combinatorial Theory (B)* **24** (1978), 14–23; *MR***58**#21808.
11. V. Chvátal, A remark on a problem of Harary, *Czech. Math. J.* **20** (1970), 109–111; *MR***42**#1694.
12. J. Chvátalová, On the bandwidth problem for graphs, Ph.D. Thesis, Department of Combinatorics and Optimization, University of Waterloo, 1980.
13. J. Chvátalová, A. K. Dewdney, N. E. Gibbs and R. R. Korfhage, The bandwidth problem for graphs: a collection of recent results, Research Report #24, Department of Computer Science, University of Western Ontario, London, Ontario, 1975.
14. J. Chvátalová and J. Opatrný, Two results on the bandwidth of graphs, *Proc. Tenth Southeastern Conf. on Combinatorics, Graph Theory and Computing I, Congressus Numerantium XXIII*, Utilitas Math., Winnipeg, 1979, pp. 263–273; *MR***81f**#05064.

15. J. Chvátalová and J. Opatrný, The bandwidth problem and operations on graphs, preprint.
16. E. Cuthill and J. McKee, Reducing the bandwidth of sparse symmetric matrices, *Proc. 24th Natl. Conf. ACM*, 1969, pp. 157–172.
17. A. K. Dewdney, Tree topology and the *NP*-completeness of tree bandwidth, Research Report no. 60, Department of Computer Science, University of Western Ontario, London, Ontario, 1980.
18. P. G. Eitner, The bandwidth of the complete multipartite graph, Presented at *Toledo Symposium on Applications of Graph Theory*, 1979.
19. M. R. Garey, R. L. Graham, D. S. Johnson and D. E. Knuth, Complexity results for bandwidth minimization, *SIAM J. Appl. Math.* **34** (1978), 477–495; *MR***57**#18220.
20. M. R. Garey and D. S. Johnson, *Computers and Intractability: A Guide to the Theory of NP-completeness*, W. H. Freeman, San Francisco, 1979; *MR***80g**:68056.
21. M. R. Garey, D. S. Johnson and R. L. Stockmeyer, Some simplified *NP*-complete graph problems, *Theoretical Comput. Sci.* **1** (1976), 237–267; *MR***53**#14978.
22. M. R. Garey, D. S. Johnson and R. L. Stockmeyer, Some simplified *NP*-complete problems, *Proc. 6th ACM Symposium on Theory of Computing*, 1974, pp. 47–63; *MR***54**#6549.
23. N. E. Gibbs, W. G. Poole, Jr. and P. K. Stockmeyer, An algorithm for reducing the bandwidth and profile of a sparse matrix, *SIAM J. Numer. Anal.* **13** (1976), 236–250; *MR***58**#19061.
24. M. K. Goldberg and I. A. Klipker, Minimal placing of trees on a line (Russian), Technical report, Physico-Technical Institute of Low Temperatures, Academy of Sciences of Ukranian SSR, USSR, 1976.
25. S. W. Golomb, How to number a graph, *Graph Theory and Computing* (ed. R. C. Read), Academic Press, New York, 1972, pp. 23–37; *MR***49**#4863.
26. E. M. Gurari and I. H. Sudborough, Improved dynamic programming algorithms for the bandwidth minimization problem and the min cut linear arrangement problem, Technical Report, Department of Electrical Engineering and Computer Science, Northwestern University, Evanston, Ill., 1982.
27. F. Harary, *Graph Theory*, Addison-Wesley, Reading, Mass., 1969; *MR***41**#1566.
28. L. H. Harper, Optimal assignments of numbers to vertices, *J. Soc. Indust. Appl. Math.* **12** (1964), 131–135; *MR***29**#41.
29. L. H. Harper, Optimal numberings and isoperimetric problems on graphs, *J. Combinatorial Theory* **1** (1966), 385–393; *MR***34**#91.
30. M. A. Iordanskiĭ, Minimal numberings of the vertices of trees, *Soviet Math. Dokl.* **15** (1974), 1311–1315; *MR***50**#9649.
31. F. T. Leighton, New lower bound techniques for VLSI, *Proc. 22nd IEEE Symposium on Foundations of Computer Science*, 1981, pp. 1–12.
32. T. Lengauer, Upper and lower bounds on the complexity of the min-cut linear arrangement problem on trees, *SIAM J. Alg. Discrete Methods* **3** (1982), 99–113; *MR***83a**:68081.
33. N. Megiddo, L. Hakimi, M. R. Garey, D. S. Johnson and C. H. Papadimitriou, The complexity of searching a graph, *Proc. 22nd IEEE Symposium on Foundations of Computer Science*, 1981, pp. 376–385.
34. F. Makedon, Layout problems and their complexity, Ph.D. Thesis, Department of Electrical Engineering and Computer Science, Northwestern University, Evanston, Ill., 1982.
35. F. Makedon, C. H. Papadimitriou and I. H. Sudborough, Topological bandwidth, preprint.
36. C. H. Papadimitriou, The *NP*-completeness of the bandwidth minimization problem, *Computing* **16** (1976), 263–270; *MR***53**#14981.
37. M. Paterson, W. Ruzzo and L. Snyder, Bounds on minimax edge length for complete binary trees, *Proc. 13th ACM Symposium on Theory of Computing*, 1981, pp. 293–299.
38. A. L. Rosenberg and I. H. Sudborough, Bandwidth and pebbling, preprint.
39. J. B. Saxe, Dynamic programming algorithms for recognizing small-bandwidth graphs in polynomial time, *SIAM J. Alg. Discrete Methods* **1** (1980), 363–369; *MR***82a**:68086.

40. M. A. Seidvasser, The optimal numbering of the vertices of a tree, *Diskret. Analiz* **17** (1970), 56–74; *MR***45**#105.
41. Y. Shiloach, A minimum linear arrangement algorithm for undirected trees, *SIAM J. Comput.* **8** (1979), 15–32; *MR***80c**:68034.
42. I. H. Sudborough and J. Turner, On computing the width and black/white pebbles demands of a tree, preprint.
43. M. M. Sysło and J. Zaks, The bandwidth problem for ordered caterpillars. Computer Science Department Report CS-80-065, Washington State University, 1980.
44. M. Yannakakis, A polynomial algorithm for the min-cut linear arrangement of trees, *Proc. 24th IEEE Symposium on Foundation of Computer Science*, 1983, pp. 274–281.

8
Polytopal Graphs

JOSEPH MALKEVITCH

1. Introduction

By the time of Euclid the study of polyhedra was already quite advanced. For example, Book XII of Euclid's *Elements* shows how to compute the volume of a pyramid (a result attributed by Archimedes to Eudoxus). The so-called *regular* or *Platonic* solids (the tetrahedron, cube, octahedron, icosahedron and dodecahedron) were presented in Book XIII; these are the solids in which each face is the same regular polygon and each vertex lies on the same number of faces. In the modern definition of a semi-regular solid, each face is again a regular polygon (not all necessarily the same), and the arrangement of faces around each vertex is the same. There are 13 such solids, and they were apparently first studied by Archimedes, although the original manuscript has been lost, and our knowledge is due to a summary of his work by Pappus.

Most of this early work was metrical in nature, including Euclid's proof that there are only five regular solids. The lovely result (*Euler's polyhedron formula*) that, in a convex polyhedron with v vertices, e edges and f faces, $v - e + f = 2$, seems to have eluded the Greek geometers. It was apparently discovered by Euler (not by Descartes who is often credited with it) and first proved by Legendre. It is interesting that Euler was unable to prove the formula, even though he looked at it from a non-metrical point of view (see, for example, Biggs *et al.* [**8**]). The graph theory version of the formula states that, for a plane connected graph

GRAPH THEORY, 3
ISBN 0–12–086203–4

with v vertices, e edges and f faces, $v - e + f = 2$; in this form, graph-theoretic proofs are not difficult to find.

Euler's formula was not systematically exploited to any extent until the late nineteenth century. Only then was there a renaissance in interest in the metrical and combinatorial properties of solids. Among the noteworthy contributors to these investigations were J. Steiner, T. P. Kirkman, V. Eberhard, M. Brückner, V. Schlegel, L. Schläfli and H. Poincaré. These mathematicians generated renewed interest in the geometry of convex solids, an interest which culminated in the landmark book *Vorlesungen über die Theorie der Polyeder* [**79**] by Steinitz and Rademacher, published in 1934. This book presented an extraordinary result, known as *Steinitz' theorem*, characterizing the combinatorial structure of convex solids. Unfortunately, the theorem was couched in such archaic language that its value was not appreciated for many years. It was only after the reformulation of the theorem by B. Grünbaum in 1963 that it released a torrent of results. This resulted in a cross-fertilization of geometry with both graph theory and combinatorics, with benefit to all three areas.

Of course, there were other factors contributing to this renewed interest in polytopes, perhaps the most important of these being linear programming. Since the region of feasibility for a linear program is a polyhedron, the polyhedra associated with specially structured linear programs (such as the transportation and assignment problems) have been studied extensively in a field known as *polyhedral combinatorics*.

The purposes of this survey are:

(*i*) to describe some important results on polytopes and their graphs;

(*ii*) to show how graphs can be used in the study of polytopes;

(*iii*) to present unsolved problems—large and small—in this area.

We have had to be selective in our choice of topics; for fuller accounts, we refer to the books by Grünbaum [**33**], Brøndsted [**13**], and Yemelichev *et al.* [**92**].

2. Definitions and Steinitz' Theorem

In this brief section, we introduce some of the geometric terminology used in the study of polytopes. Our discussion will be confined to $\mathbf{E}^d$ (d-dimensional Euclidean space), with most emphasis on $d = 3$. All graphs are assumed simple unless otherwise specified.

A set X in $\mathbf{E}^d$ is **convex** if the segment joining any pair of points in X also lies in the set. The **convex hull** of a set S is the intersection of all convex sets containing S. A ***d*-polytope** is the convex hull of some finite set of points which has dimension d. (Polytopes can also be defined using intersections of half-spaces; for a discussion of the connection between the two approaches, see Brøndsted [**13**].)

A d-polytope has a 'face structure', with faces of dimensions 0 to $d - 1$. All we shall need are those of dimensions 0, 1 and $d - 1$, called respectively the **vertices**, **edges** and **facets** (**faces** if $d = 3$) of the polytope. The **graph of a polytope** P is the graph consisting of the vertices and edges of P. A graph G is called ***d*-polytopal** if it is isomorphic to the graph of some d-polytope P; we then say that P **realizes** G.

We now consider characterizations of d-polytopal graphs for $d \leq 3$. Since 1-polytopes are just line segments, the only 1-polytopal graph is the complete graph K_2. Further, every 2-polytope is a convex polygon, and so a graph is 2-polytopal if and only if it is a circuit C_p, for some $p \geq 3$. Note that, for $p > 3$, C_p can also be realized by simple non-convex polygons in the plane, and these too have an interesting theory. More generally, a d-polytopal graph may also be realized by a 'non-convex d-polytope'. In particular, Grünbaum [**38**] has indicated that a flat-faced non-convex solid may have a complete graph as its graph; for a survey of this surprisingly rich theory, see Grünbaum [**37**].

This brings us to the case $d = 3$ and to what could be called the 'Fundamental Theorem of Polytopal Graph Theory'—namely, **Steinitz' theorem** characterizing 3-polytopal graphs. In modern terminology, its statement is most elegant:

Theorem 2.1 (Steinitz' Theorem). *A graph is 3-polytopal if and only if it is planar and 3-connected.* ||

This result is deeper than it might at first appear; proofs and excellent discussions may be found in [**5**], [**33**], [**40**], [**56**] and [**92**]. What makes the theorem so remarkable is its implication that a general class of 3-dimensional structures is equivalent to a certain class of 2-dimensional ones—that is, studying 3-polytopes combinatorially does not require thinking of them in 3-dimensional space. In the next section, we take advantage of some of the power of this result in an analysis of some existence questions.

We conclude this section with a couple of comments. First, the connection between 3-polytopes and Schlegel diagrams is still unresolved; we refer the reader to Grünbaum [**33**], [**36**], for more on this topic. Secondly, the connections among embedding graphs in the plane, 3-connectivity and convex representations of planar graphs have been given by Thomassen [**80**], [**82**], [**83**], [**85**], [**86**], and by Grünbaum and Shephard [**41**].

3. Face Sizes in Regular 3-Polytopes

In this section and the next we consider questions related to the number of sides that the various faces of k-valent 3-polytopes can have. For convenience, we shall omit the prefix "3-" in the next few sections, so that "polytope" will always mean "3-polytope".

An elementary consequence of Euler's polyhedron formula states that every planar graph has a vertex of degree no greater than 5; thus, the only regular polytopal graphs are 3-valent, 4-valent or 5-valent. The following result gives some relationships among the numbers of faces of various sizes in these graphs:

Theorem 3.1. *Let G be an r-valent polytopal graph having p_k faces with k sides ($k \geq 3$). Then*

(*i*) *for* $r = 3$, $\quad 3p_3 + 2p_4 + p_5 = 12 + \sum_{k \geqslant 6} (k - 6)p_k$;

(*ii*) *for* $r = 4$, $\quad p_3 = 8 + \sum_{k \geqslant 4} (k - 4)p_k$;

(*iii*) *for* $r = 5$, $\quad p_3 = 20 + \sum_{k \geqslant 4} (3k - 10)p_k$.

Proof. If G has v vertices, e edges and f faces, then by Euler's formula and simple counting, we have

$$v - e + f = 2, \quad f = \sum_{k \geqslant 3} p_k \text{ and } 2e = \sum_{k \geqslant 3} kp_k.$$

Furthermore, in an r-valent graph, $rv = 2e$, and (*i*), (*ii*) and (*iii*) follow from these relationships for the corresponding values of r. ||

It is not difficult to see that none of these three equations is sufficient for the existence of an r-valent polytope with p_k faces of size k. For example, there is no trivalent polytope with $p_3 = 4$ and $p_6 = 1$ (and all other $p_k = 0$), no 4-valent polytope with $p_3 = 8$ and $p_4 = 1$, and no 5-valent polytope with $p_3 = 22$ and $p_4 = 1$, even though these numbers satisfy the above equations. It is therefore natural to ask what additional conditions must be satisfied so as to guarantee the existence of a polytope with a given number of faces of each size.

It is convenient to introduce further notation and terminology. Given a polytope P, we let $p_k(P)$ (or, simply, p_k) denote the number of faces with k sides, and let $\pi(P)$ denote the sequence $\{p_3, p_4, \ldots\}$. A sequence $\pi^* = \{p_3^*, p_4^*, \ldots\}$ of non-negative integers is said to be **realizable** if $\pi(P) = \pi^*$, and P is said to **realize**, or **be a realization of**, π^*.

Our first result follows from work on 2-connected plane graphs by Grünbaum [**33**] (see also [**36**] for generalizations), who showed that the face numbers of such graphs satisfy congruences similar to the equations of Theorem 3.1. In addition, he proved, for $m = 2, 3, 4$ or 5, that there is no 2-connected plane graph in which the number of sides of every face except one is a multiple of m. Furthermore, if there are precisely two exceptional faces, then they cannot share an edge. The following theorem is a consequence of those results:

Theorem 3.2. *If $\pi = \{p_3, p_4, \ldots\}$ is realizable by a trivalent polytope, then, for $m = 2, 3, 4$ and 5,*

$$\sum_{m \nmid k} p_k \neq 1. \parallel$$

Trivalent polytopes with just two types of faces were investigated by Malkevitch [**60**]. Among his results is the following:

Theorem 3.3. *If a trivalent polytope has only triangles and k-gons as faces (for one value of k), then $k \leq 10$.*

Proof. It is a consequence of Steinitz' theorem that if a trivalent polytope contains two triangles with a common side, then it is a tetrahedron. Therefore, in any trivalent polytope with more than four vertices, no two triangles can even have a vertex in common. Suppose that, for some $k \geq 11$, there is a polytope P with only triangles and k-gons as faces. Then, contracting each triangle to a point (a new vertex), we get a plane graph in which every face is at least a hexagon, which is impossible. ||

Turning now to the 4-valent case, we have the following result of Enns [**18**], which he showed is best possible:

Theorem 3.4. *If π^* satisfies equation (ii) of Theorem 3.1 and if $p_4^* \geq p_3^*$, then π^* is realizable by a 4-valent polytope.* ||

Enns has also considered the 4-valent case in which digons are allowed (and hence the graphs are no longer polytopal).

The realizability of sequences by 5-valent polytopes was considered by Fisher [**25**]; his results have been improved by Trenkler [**87**].

Other results which shed light on questions of polytopal realizability can be found in papers on degree-sequences of planar graphs. If a degree-sequence $(\rho_1, \rho_2, \ldots, \rho_p)$ has a plane realization, then the faces of the dual graph have sizes ρ_i, from which the values of p_k can be deduced. For results in this area, see Chvátal [**15**], Farrell [**22**], Jendrol' [**52**], and Schmeichel and Hakimi [**76**].

In the next section we look at one particular aspect of realizability in greater detail. Before getting to that, we have a few other comments to make. Many mathematicians are interested not only in whether a certain type of structure exists, but in how many such structures there are. By a variety of techniques—theoretical concepts, hand calculations and computer methods—many questions on the number of polytopes of a certain type (such as having f faces or e edges) have been explored. In many cases, there are exact counts for small values. In these counts, only combinatorially distinct polytopes are usually considered different, but sometimes they are considered to be 'rooted' at a face or an edge. One outstanding unsolved problem is whether the automorphism group of almost all polytopes is asymptotically the identity group. The reader who wishes to study enumeration questions further may find the following references useful: for theoretical approaches, see Tutte [**90**]; for computer calculations, see Engel [**17**]; and for a combination, see Federico [**23**] and Grünbaum [**36**].

The 'nets' of polytopes constitute an interesting topic (to which graph theory may contribute) which unfortunately has been relegated in large part to the area of recreational mathematics. On an intuitive level, a *net* is obtained from a polytope by cutting some of its edges and 'opening up the polytope' so as to form a planar diagram. Figure 1 shows two nets of a cube.

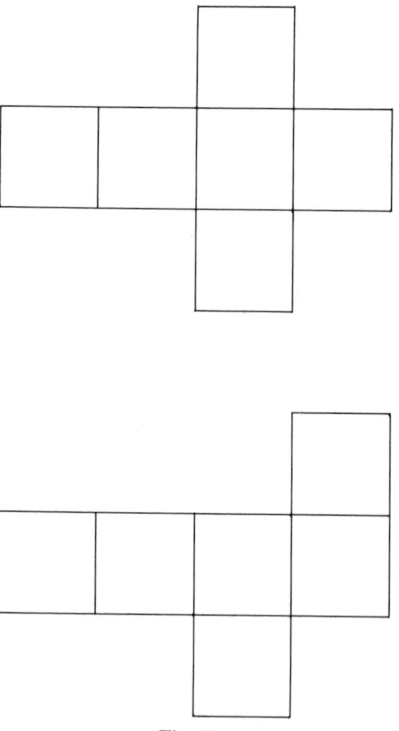

Fig. 1

By reversing the procedure—that is, by pasting along the edges common to polygons, and then taping joins—one can recover the polytope, or perhaps even form a new one. In fact, non-isomorphic polytopes can give the same net, so in general neither the polytope nor the net necessarily determines the other. Hence, one must be more formal about the definition of a net than we have been here. Problems on the enumeration of nets and the selection of nets with special properties are of interest when building models of polytopes. For precise definitions, basic concepts and far-ranging results on nets, see the important paper of Shephard **[77]**.

4. Eberhard's Problem

The form of part (*i*) of Theorem 3.1 suggests a degree of freedom, due to the fact that the coefficient of p_6 is 0. Specifically, if we have available a collection of non-hexagonal polygons whose numbers satisfy (*i*), can we construct a trivalent polytope by including a suitable number of hexagons? These hexagons can be thought of as a 'free supply of fillers' to be used in case trouble arises during the construction of a polytope from the given non-hexagonal supply. This question was 'answered' by V. Eberhard (see Grünbaum **[33]**, **[35]**).

Before stating his result, we introduce some notation which will be useful throughout this discussion: we let $\sigma^* = (p_3^*, p_4^*, p_5^*, p_7^*, p_8^*, \ldots)$ denote a sequence of non-negative integers—note the absence of p_6^*. (We also assume, without loss of generality, that only a finite number of these terms are positive.) We say that σ^* is **augmentable by p_6^*** if there exists a trivalent polytope P with $p_k(P) = p_k^*$ for all $k \geq 3$—that is, P is a realization of σ^* augmented by p_6^*. Note that p_6^* is allowed to be zero.

Theorem 4.1. *Every sequence σ^* which satisfies Theorem* 3.1(*i*) *is augmentable by some p_6^*.* ||

The known proofs of this intriguing and elegant result are all rather detailed and involve careful book-keeping (see, for example, Grünbaum **[33]** or Jendrol' **[52]**). Some remarkable generalizations have been found by Horňák **[51]** and Jendrol' **[53]**.

Various bounds on the value of p_6^* in Eberhard's theorem have been found. A particularly nice result of Grünbaum **[34]** states that, if $p_3^* = p_4^* = 0$, then $p_6^* \leq 8$. On the other hand, some sequences require p_6^* to be large; for example, (p_3^*, p_k^*) for $k \geq 12$ (whence (*i*) becomes $p_3^* = 12 + (k-6)p_k^*$). Other bounds have been obtained by Barnette **[3]** and Jucovič **[57]**. In particular, Barnette has shown that

$$p_6^* \geq 2 + \tfrac{1}{2}p_3^* - \tfrac{1}{2}p_5^* - \sum_{k\geqslant 7} p_k^*.$$

The following question related to Eberhard's theorem has attracted considerable attention:

if $3p_3^* + 2p_4^* + p_5^* = 12$ (*that is*, $p_k^* = 0$ *for* $k \geq 7$), *for which* p_6^* *is the sequence* $(p_3^*, p_4^*, p_5^*, p_6^*)$ *realizable by a trivalent polytope?*

This particular problem has also been of interest to geologists, since faces of polyhedra occurring in nature rarely have more than six sides. Partial solutions have been found by various mathematicians, and many results have been rediscovered; for a complete solution, see Grünbaum [**34**] and Malkevitch [**61**].

The general problem of determining all admissible values of p_6^* in Eberhard's theorem was treated by Fisher [**24**]. We include one of his theorems here; it is quite striking, but the original version required a small correction:

Theorem 4.2. *Let* σ^* *be a solution of Theorem* 3.1(*i*), *in which either* $p_4^* \geq 2$ *or* $p_5^* \geq 2$, *and let* S *denote the sum of the entries in* σ^*. *Then there exists an integer* $M \leq 3S$ *such that, for all* $p_6^* \geq M$, σ^* *is augmentable by* p_6^*. ||

Along similar lines, the following theorem was discovered by Jendrol' [**54**]:

Theorem 4.3. *Let* σ^* *be a solution of Theorem* 3.1(*i*) *which has total sum* S, *and let* $T = S - \sum_{j \geq 3} p_j$. *Then*

(*i*) *if* $T \leq 2$ *and* S *is even, there exists an integer* M *such that* σ^* *is augmentable by every even integer* $p_6^* \geq M$ *and by no odd integer*;

(*ii*) *if* $T \leq 2$ *and* S *is odd, there exists an integer* M *such that* σ^* *is augmentable by every odd integer* $p_6^* \geq M$ *and by no even integer*;

(*iii*) *if* $T \geq 3$, *there exists an integer* M *such that* σ^* *is augmentable by every integer* $p_6^* \geq M$. ||

Jendrol' conjectured that in all cases, the minimum value of M does not exceed $3S$.

In the 4-valent case, it is the coefficient of p_4 which is 0, so the 'filler' consists of 4-gons. We let τ^* denote a sequence $(p_3^*, p_5^*, p_6^*, \ldots)$, and we say that τ^* is **augmentable by** $\boldsymbol{p_4^*}$ if the inclusion of p_4^* yields a sequence which is realizable by some 4-valent polytope. The analogue of Eberhard's theorem was found by Grünbaum [**33**]:

Theorem 4.4. *Every sequence* τ^* *which satisfies Theorem* 3.1(*ii*) *is augmentable by some* p_4^*. ||

A proof which is delightfully easy to follow can be found in Grünbaum

[**33**]; other proofs have been given by Jucovič [**58**] and Malkevitch [**64**]. Grünbaum's original proof deserves to be better known, and so we illustrate it here with an example.

In general procedure is to construct a 'block' for each k-gon with $k \geq 5$, consisting of $k - 4$ triangles, one k-gon and some 'free' 4-gons. For our example, we take $p_3^* = 13$, $p_6^* = p_7^* = 1$ (and all other p_k^* are 0, except p_4^*). For a 6-gon and a 7-gon, these blocks are shown in Fig. 2(a). We then arrange the blocks in a diagonal fashion and form a square pattern using more 4-gons, as in Fig. 2(b). At this stage in our example there are just five triangles. Additional vertices and edges are then added to form a graph which is 4-valent and has eight more triangles (see Fig. 2(c)). The resulting graph is planar and 3-connected and so, by Steinitz' theorem, is the desired realization.

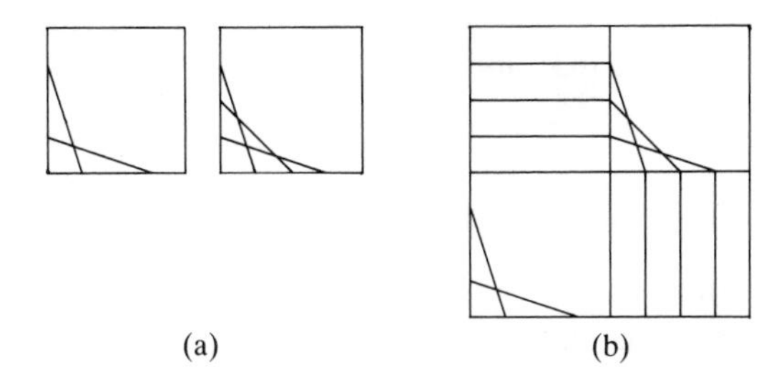

(a) (b)

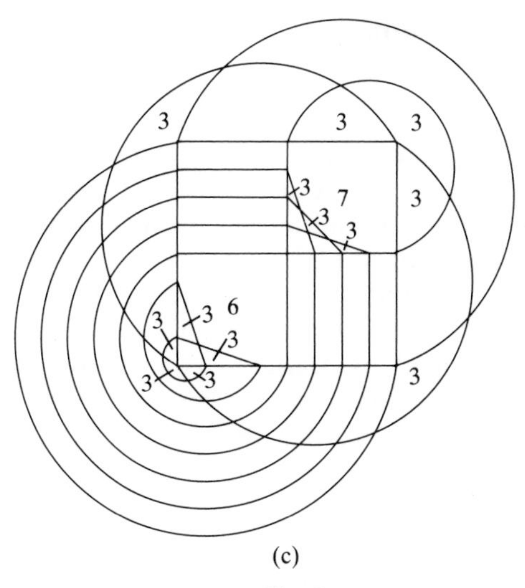

(c)

Fig. 2

Enns' theorem (Theorem 3.4) is a strengthening of Grünbaum's result, in that it states that such a realization exists for every $p_4^* \geq p_3^*$. A further improvement has been given by Trenkler [**88**]. Note that, since no coefficient is 0 in Theorem 3.1(*iii*), there is no 5-valent version of Eberhard's theorem.

Another 'Eberhard-type problem' concerns the presence of 2-valent vertices in spanning trees of regular polytopal graphs. This area of investigation arose from the observation of a zero coefficient in a degree equation for trees:

Theorem 4.5. *If T is a non-trivial tree with t_k vertices of degree k, then*

$$t_1 = 2 + \sum_{k \geqslant 2} (k - 2)t_k. \quad \| \qquad (*)$$

The following analogue of Eberhard's theorem was found by Malkevitch [**63**]; further work on degrees of spanning trees in polytopal graphs can be found in Malkevitch [**63**] and Joffe [**56**]:

Theorem 4.6. *For $r = 3, 4$ or 5, if $t_1^*, t_3^*, t_4^*, \ldots$ is a solution of $(*)$ in non-negative integers, then there exists an integer t_2^* and an r-valent polytope P such that P has a spanning tree with t_k^* vertices of degree k.* ||

We conclude this section by noting that extensive generalizations of Eberhard's theorem to surfaces of higher genus are also known (see Grünbaum [**36**]).

5. Hamiltonian Problems: Trivalent Polytopes

The development of the theory of Hamiltonian circuits in planar graphs was spurred by a connection with the four-color problem. If every trivalent polytopal graph could be proved to be Hamiltonian, then the four-color theorem could be deduced. A discussion of the connections between the four-color problem and Hamiltonian circuits can be found in Biggs *et al.* [**8**] and Saaty and Kainen [**75**].

The first example of a non-Hamiltonian trivalent polytope was found by W. T. Tutte in 1946; it has 46 vertices and is shown in Fig. 3. Other examples have been found by Barnette, Lederberg, Bosák, Hunter, Faulkner and Younger, with one example having just 38 vertices. Holton and McKay [**48**] recently showed that there is none with fewer vertices:

Theorem 5.1. *The fewest vertices in any non-Hamiltonian* 3*-valent polytope is* 38. ||

A major breakthrough in speeding the search for non-Hamiltonian

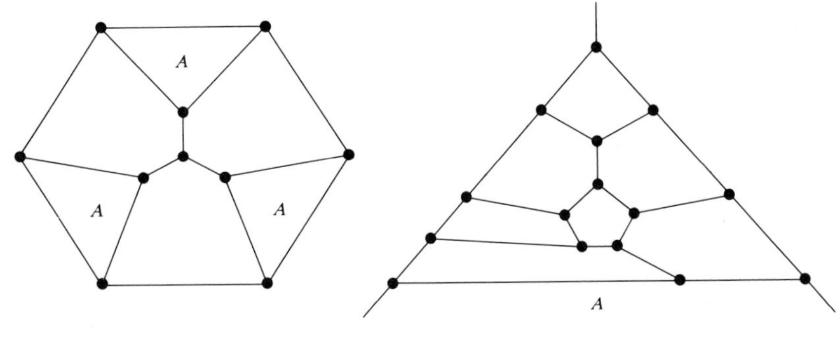

Fig. 3

polytopal graphs was the so-called **Grinberg condition**; in retrospect, it seems surprising that this result was not discovered earlier:

Theorem 5.2. *Let G be a plane graph of order p having a Hamiltonian circuit C, and let q_k and r_k denote the numbers of faces with k sides which lie inside and outside C, respectively. Then*

$$\sum_{k\geqslant 3} (k-2)(q_k - r_k) = 0. \qquad (\dagger)$$

Proof. Let d denote the number of edges inside C. Then

$$\sum_{k\geqslant 3} kq_k = 2d + p.$$

It is easily seen that $\sum_{k\geqslant 3} q_k$, the total number of faces inside C, equals $d + 1$. These two relationships imply that $\sum_{k\geqslant 3} (k-2)q_k = p - 2$. Applying the same argument to the edges outside C gives $\sum_{k\geqslant 3} (k-2)r_k = p - 2$, and the theorem follows. ||

The use of (†) as a powerful tool for constructing non-Hamiltonian graphs works this way: if all faces but one in a trivalent polytopal graph have sizes congruent to 2 (modulo 3), then (†) cannot hold and H must be non-Hamiltonian. A graph is called **cyclically *k*-edge-connected** if G cannot be separated into two components, both containing circuits, by removing fewer than k edges. Grinberg's condition was useful in finding graphs of least order which are cyclically k-edge-connected for $k = 3$, 4 and 5; these graphs may be found in **[29]**, **[14]** and **[50]**. More on the Grinberg condition appears in Honsberger **[50]**, and extensions can be found in **[9]**, **[28]** and **[90]**.

Spurred by suggestions that the Grinberg condition was the key to showing non-Hamiltonicity, Zaks [**95**] constructed an example of a non-Hamiltonian graph which is also 'non-Grinbergian'—that is, a trivalent polytopal non-Hamiltonian graph in which each face is either a 5-gon or an 8-gon. In fact, he showed much more than this:

Theorem 5.3. *For each $k \geq 8$, there exists a non-Hamiltonian 3-polytopal trivalent graph in which every face is a 5-gon or a k-gon.* ||

Owens [**66**] has extended these results to other families in which there are only one or two types of faces.

Results on those trivalent polytopal graphs which have Hamiltonian circuits are very meager. One exception to that statement is the following result of Goodey [**29**], [**30**]:

Theorem 5.4. *Every trivalent polytopal graph in which all faces are either*

(*i*) *triangles or hexagons, or*

(*ii*) *4-gons or hexagons*

is Hamiltonian. ||

This result raises an interesting open question, posed by D. Barnette:

if no face in a trivalent polytopal graph has more than six sides, must the graph have a Hamiltonian circuit?

Another open conjecture of relatively long standing was posed by D. Barnette:

is every bipartite trivalent polytopal graph Hamiltonian?

It is known (see Holton *et al.* [**47**]) that any counter-example to Barnette's conjecture must have at least 64 vertices. An alternative way of stating the conjecture is that if the number of sides of every face of a trivalent polytope is a multiple of 2, then the polytope is Hamiltonian. The special case in which 2 is replaced by 4 is no easier to establish. Non-Hamiltonian trivalent polytopes are known in which the number of sides of each face is a multiple of 3 and in which all are a multiple of 5. Thus, there is circumstantial evidence that Barnette's conjecture may be false. Related problems are discussed in [**47**] and [**67**].

Not only is the Grinberg condition useful in constructing non-Hamiltonian graphs, but it can also be used to construct Hamiltonian graphs in which there are edges which lie on every Hamiltonian circuit or on none (see, for example, Bosák [**12**]). Other interesting questions arise when one studies polytopes with a fixed number of Hamiltonian circuits, or in which each edge lies on a given number of them.

A graph is called **hypo-Hamiltonian** if it is not Hamiltonian but each of its vertex-deleted subgraphs is. There had been speculation that there were no trivalent hypo-Hamiltonian polytopes, but Thomassen **[81]** recently showed that such graphs do in fact exist (see Fig. 4).

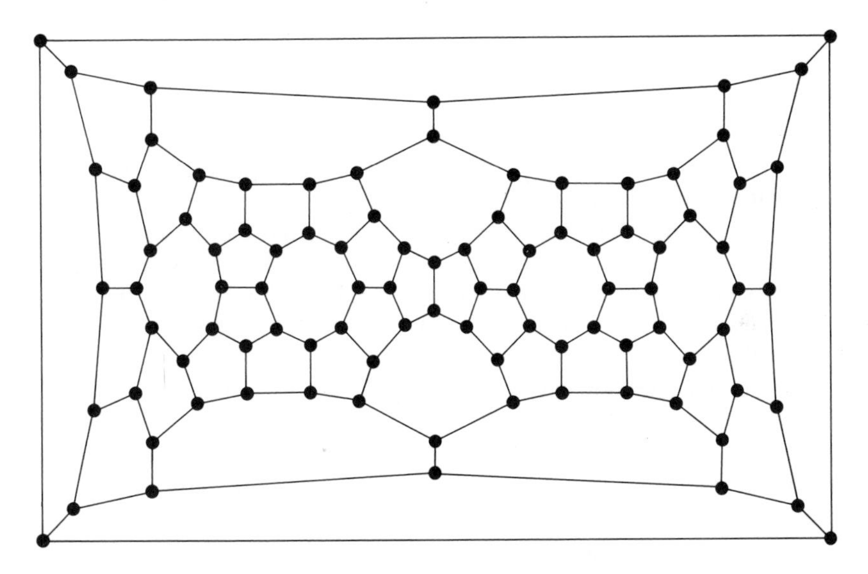

Fig. 4

6. Hamiltonian Problems: Other Aspects

There are many interesting problems about Hamiltonian circuits in polytopes other than trivalent ones, and we now consider some of these. By far the most important result is the following beautiful theorem of Tutte **[89]**:

Theorem 6.1. *Every planar* 4*-connected graph is Hamiltonian.* ||

In contrast, there are examples of 4-valent and 5-valent polytopes which are not Hamiltonian (see Grünbaum **[33]** and Zamfirescu **[98]**).

There is also considerable interest in plane triangulations; all are polytopal (if their order is at least 4), but not all are Hamiltonian. For example, if in a plane triangulation with five vertices, we erect a pyramid on each face (add a vertex of degree 3 inside each face and join it to the face's vertices), then the result is a non-Hamiltonian triangulation since no two of the six new vertices are adjacent, so there could be no circuit containing all 11 vertices (see also **[14]** or **[33]**). There are, however, some positive statements which can be made about plane triangulations. One is an old result due to Whitney [**91**], generalized by Tutte's theorem; another is a relatively new result of Ewald [**20**]:

Theorem 6.2. *Let G be a plane triangulation of order at least* 4. *If either*

(*i*) *every triangle of G bounds a face, or*

(*ii*) *no vertex of G has degree greater than* 6,

then G is Hamiltonian. ||

For other work on circuits and paths in plane triangulations, see Hakimi and Schmeichel [**45**], [**46**].

Grünbaum and Malkevitch [**39**] have investigated the existence of pairs of edge-disjoint Hamiltonian circuits in 4-valent Hamiltonian polytopes, and Zaks [**93**] has treated the 5-valent case. In both instances, there need not be such a pair of disjoint circuits; for more on this topic, see also Bielig [**7**] and Zamfirescu [**97**], [**98**].

Considerable attention has been given to algorithms for finding Hamiltonian circuits. Because of the extensive literature on the subject, we note here only the result of Garey *et al.* [**27**] that the problem of finding Hamiltonian circuits in polytopes is *NP*-complete, and that Plesnik showed this to be true for just bipartite trivalent polytopes.

We turn now to questions involving lengths of circuits other than Hamiltonian ones. Some definitions are in order. A graph of order p is **pancyclic** if it has a circuit of each length k for $k = 3, 4, \ldots, p$, and is **almost pancyclic** if it is Hamiltonian and has a circuit of each length k for $k = 3, 4, \ldots, p - 1$, except one.

Consider a graph formed in the following way: take a tree T of order at least 4 and having no vertices of degree 2, and embed it in the plane; then join its end-vertices sequentially with a polygon enclosing T (see Fig. 5). The result is called a **Halin graph** and is always polytopal. That such graphs also have circuits of many lengths was shown independently by Bondy and Lovász [**10**] and Skowrońska [**78**]:

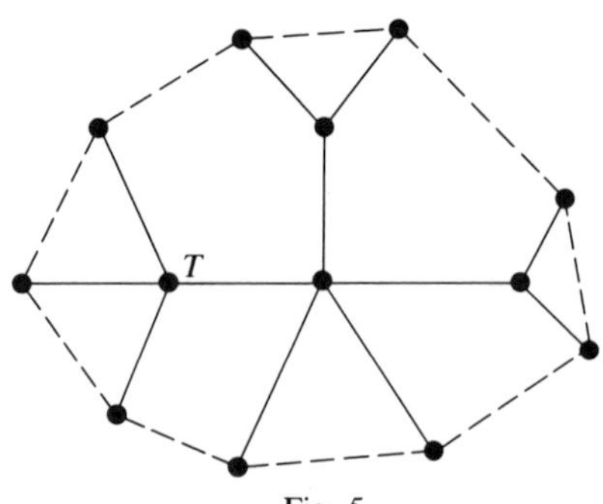

Fig. 5

Theorem 6.3. *Every Halin graph is either pancyclic or almost pancyclic with the missing circuit length even.* ||

For general Halin graphs, this result is best possible, but Bondy and Lovász used the same methods to prove the following:

Corollary 6.4. *If the tree used in forming a Halin graph has no vertices of degree* 3, *then the Halin graph is pancyclic.* ||

Another problem involving pancyclicity arises out of Tutte's theorem:

Conjecture 6.1. *If a* 4-*connected polytope has a circuit of length* 4, *then it is pancyclic.*

Along these lines, Thomassen [**84**] verified a conjecture of M. D. Plummer:

Theorem 6.5. *Every* 4-*connected polytopal graph has a Hamiltonian path between each pair of vertices.* ||

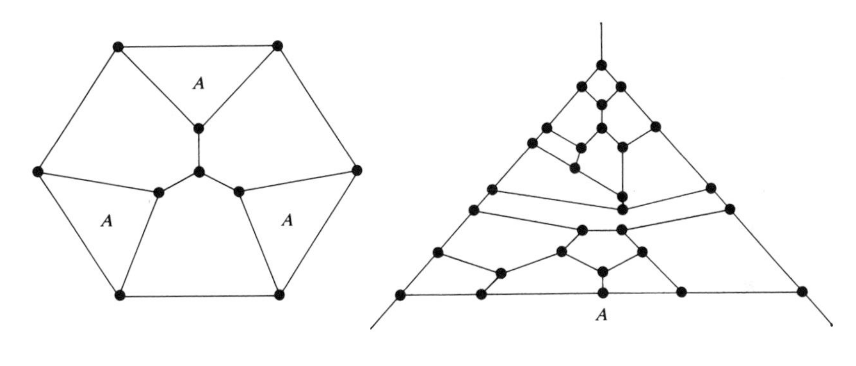

Fig. 6

The smallest known 3-valent polytopal graph which has no Hamiltonian path is due to Zamfirescu [**97**], [**98**] and is shown in Fig. 6. Other work related to paths and circuits of various lengths may be found in [**11**], [**49**] and [**71**].

A further vein of interesting work arises from the concept of the *shortness exponent* of a family of graphs. The basic idea is this: in a family of non-Hamiltonian graphs, how does the length of the longest circuit in each graph compare with the order of the graph? This shortness exponent thus measures how 'non-Hamiltonian' a family can become; see Ewald [**21**], Grünbaum and Walther [**42**] and Owens [**66**] for details.

7. Polytopes of Dimension Greater than 3

Graphical problems of higher-dimensional polytopes put one in virtually uncharted waters. It is perhaps too much to hope that, for $d \geq 4$, only one graph can capture the structure of a d-polytope. Evidence for this is provided by the following theorem:

Theorem 7.1. (*i*) *For $d \geq 4$, the complete graph K_p is d-polytopal for all $p \geq d + 1$;*

(*ii*) *given a graph G, there can exist two d-polytopes ($d \geqslant 4$) P_1 and P_2, which are not isomorphic, but whose graphs are both isomorphic to G.* ||

These facts mean that, for a higher-dimensional polytope, the graph of its vertices and edges does not capture the facet structure, nor even the dimension of the polytope. The following theorem contains some facts about d-polytopal graphs; the first result is due to Balinski and the second is due to Grünbaum and Motzkin [**40**]:

Theorem 7.2. *Every d-polytopal graph is d-connected and contains a subdivision of K_{d+1} as a subgraph.* ||

As with 3-polytopes, the question of the existence of Hamiltonian circuits is of considerable interest. In particular, Hamiltonian circuits of d-dimensional cubes always exist, and are of interest in applications to error-correcting codes and analog-to-digital conversion. The latter application arises if, on a circular dial, 2^d sectors are to be labeled with d binary digits so as to minimize the number of digit changes in going from each sector to the next. Any Hamiltonian circuit on the d-cube gives a solution with only one digit change at each stage. This problem has led to the study of 'snakes' in cubes; see, for example, Klee [**59**].

A construction similar to the one used in Section 6 for non-Hamiltonian triangulations can be used to construct non-Hamiltonian higher-dimensional polytopes. Take a d-polytope P with more facets than vertices. It can be shown that a point can be selected above each facet of P so that the convex hull P^* (known as a *Kleetope*) of these new points, together with the vertices of P, is a d-polytope in which no two new vertices are adjacent. Since there are more new vertices than old ones, it follows that P^* cannot have a Hamiltonian circuit.

We conclude with three interesting questions on Hamiltonian circuits in higher-dimensional polytopes; the first question is discussed in [**73**], [**36**], and the second is discussed in [**32**], [**74**]:

(*i*) *is every 4-valent 4-polytope Hamiltonian?*

(*ii*) *which d-dimensional prisms are Hamiltonian?*

(*iii*) *which simplicial d-polytopes are Hamiltonian?*

References

1. H. L. Abbott and P. F. Dierker, Snakes in powers of complete graphs, *SIAM J. Appl. Math.* **32** (1977), 347–355; *MR***55**#168.
2. D. W. Barnette, Trees in polyhedral graphs, *Canad. J. Math.* **18** (1966), 731–736; *MR***33**#3951.
3. D. W. Barnette, On *p*-vectors of 3-polytopes, *J. Combinatorial Theory* **7** (1969), 99–103; *MR***39**#6165.
4. D. W. Barnette, A completely unambiguous 5-polyhedral graph, *J. Combinatorial Theory* **9** (1970), 44–53; *MR***42**#5850.
5. D. W. Barnette and B. Grünbaum, On Steinitz' theorem concerning convex 3-polytopes and on some properties of planar graphs, *The Many Facets of Graph Theory* (ed. G. Chartrand and S. F. Kapoor), Lecture Notes in Math. **110**, Springer, Berlin, 1969, pp. 27–40; *MR***40**#4148.
6. D. Barnette and B. Grünbaum, Preassigning the shape of a face, *Pacific J. Math.* **32** (1970), 299–306; *MR***41**#4377.
7. G. Bielig, Untersuchungen über maximale Kreise auf 3-polytopen Graphen, Dissertation, Ruhr-Universität, Bochum, 1976.
8. N. L. Biggs, E. K. Lloyd and R. J. Wilson, *Graph Theory 1736–1936*, Clarendon Press, Oxford, 1976; *MR***56**#2771.
9. J. A. Bondy and R. Häggkvist, Edge-disjoint Hamiltonian cycles in 4-regular planar graphs, Research Report CORR 78–33, University of Waterloo, Waterloo, Ontario, Canada, 1978.
10. J. A. Bondy and L. Lovász, to appear.
11. J. A. Bondy and M. Simonovits, Longest cycles in 3-connected 3-regular graphs, Research Report CORR 78–34, University of Waterloo, Waterloo, Ontario, Canada, 1978.
12. J. Bosák, Hamiltonian lines in cubic graphs, *Theory of Graphs* (Proc. Int. Symp., Rome, 1966) (ed. P. Rosenstiehl), Dunod, Paris, 1967, pp. 35–46; *MR***36**#5022.
13. A. Brøndsted, *An Introduction to Convex Polytopes*, Graduate Texts in Mathematics **90**, Springer, New York, 1983; *MR***84d**:52009.
14. M. Capobianco and J. C. Molluzzo, *Examples and Counterexamples in Graph Theory*, North-Holland, New York, 1978; *MR***58**#10536.
15. V. Chvátal, Planarity of graphs with given degrees of vertices, *Nieuw Arch. Wisk. (3)* **17** (1969), 47–60; *MR***42**#104.
16. H. Crapo, Structural rigidity, *Structural Topology* **1** (1979), 26–45, 73; *MR***82g**:51028.
17. P. Engel, On the enumeration of polyhedra, *Discrete Math.* **41** (1982), 215–218; *MR***84h**:52001.
18. T. C. Enns, Convex 4-valent polytopes, *Discrete Math.* **30** (1980), 227–234; *MR***83a**:52012.
19. T. C. Enns, 4-valent graphs, *J. Graph Theory* **6** (1982), 255–281; *MR***83m**:05055.
20. G. Ewald, Hamiltonian circuits in simplicial complexes, *Geometriae Dedicata* **2** (1973), 115–125; *MR***47**#9424.
21. G. Ewald, On shortness exponents of families of graphs, *Israel J. Math.* **16** (1973), 53–61; *MR***49**#8892.
22. E. J. Farrell, On graphical partitions and planarity, *Discrete Math.* **18** (1977), 149–153; *MR***58**#21743.
23. P. J. Federico, Polyhedra with 4 to 8 faces, *Geometriae Dedicata* **3** (1974/5), 469–481; *MR***51**#6585.
24. J. C. Fisher, An existence theorem for simple convex polyhedra, *Discrete Math.* **7** (1974), 75–97.
25. J. C. Fisher, Five-valent convex polyhedra with prescribed faces, *J. Combinatorial Theory (A)* **18** (1975), 1–11; *MR***51**#1612.
26. S. Gallivan, Disjoint edge paths between given vertices of a convex polytope, *J. Combinatorial Theory (A)* **39** (1985), 112–115; *MR***86f**:52013.
27. M. R. Garey, D. S. Johnson and R. E. Tarjan, The planar Hamiltonian circuit problem is *NP*-complete, *SIAM J. Comput.* **5** (1976), 704–714; *MR***56**#2867.

28. K. R. Gehner, A necessary condition for the existence of a circuit of any specified length, *Networks* **6** (1976), 131–138; *MR*53#10645.
29. P. R. Goodey, Hamiltonian circuits on simple 3-polytopes, *J. London Math. Soc. (2)* **5** (1972), 504–510; *MR*47#4850.
30. P. R. Goodey, Hamiltonian circuits in polytopes with even sided faces, *Israel J. Math.* **22** (1975), 52–56; *MR*53#14314.
31. P. R. Goodey, A class of Hamiltonian polytopes, *J. Graph Theory* **1** (1977), 181–185; *MR*57#1282.
32. P. R. Goodey and M. Rosenfeld, Hamiltonian circuits in prisms over certain simple 3-polytopes, *Discrete Math.* **21** (1978), 229–235; *MR***80e**:05076.
33. B. Grünbaum, *Convex Polytopes*, Interscience, New York, 1967; *MR*37#2085.
34. B. Grünbaum, Some analogues of Eberhard's theorem on convex polytopes, *Israel J. Math.* **6** (1968), 398–411; *MR*39#6168.
35. B. Grünbaum, Planar maps with prescribed types of vertices and faces, *Mathematika* **16** (1969), 28–36; *MR*39#6768.
36. B. Grünbaum, Polytopes, graphs and complexes, *Bull. Amer. Math. Soc.* **76** (1970), 1131–1201; *MR*42#959.
37. B. Grünbaum, Polygons, *Geometry of Metric and Linear Spaces* (ed. L. M. Kelly), Lecture Notes in Math. **490**, Springer, Berlin, 1975, pp. 147–184.
38. B. Grünbaum, Polytopal graphs, *Studies in Graph Theory, II* (ed. D. R. Fulkerson), Math. Assoc. of America, Washington DC, 1975, pp. 201–224; *MR*53#10654.
39. B. Grünbaum and J. Malkevitch, Pairs of edge-disjoint Hamiltonian circuits, *Aequat. Math.* **14** (1976), 191–196; *MR*54#2544b.
40. B. Grünbaum and T. S. Motzkin, On polyhedral graphs, *Proc. Symp. Pure Math. VII*, Amer. Math. Soc., Providence, Rhode Island, 1963, pp. 285–290; *MR*27#2976.
41. B. Grünbaum and G. C. Shephard, The geometry of planar graphs, *Combinatorics: Proc. 8th British Combinatorial Conference* (ed. H. N. V. Temperley), London Math. Soc. Lecture Notes **52**, Cambridge University Press, Cambridge, 1981.
42. B. Grünbaum and H. Walther, Shortness exponents of families of graphs, *J. Combinatorial Theory (A)* **14** (1973), 364–385; *MR*47#3242.
43. B. Grünbaum and J. Zaks, The existence of certain planar maps, *Discrete Math.* **10** (1974), 93–115; *MR*50#1949.
44. S. L. Hakimi and E. F. Schmeichel, On the connectivity of maximal planar graphs, *J. Graph Theory* **2** (1978), 307–314; *MR***80a**: 05135.
45. S. L. Hakimi and E. F. Schmeichel, Graphs and their degree sequences: a survey, *Theory and Applications of Graphs* (ed. Y. Alavi and D. R. Lick), Lecture Notes in Math. **642**, Springer, Berlin, New York, 1978, pp. 225–235; *MR*58#21847.
46. S. L. Hakimi and E. F. Schmeichel, On the number of cycles of length k in a maximal planar graph, *J. Graph Theory* **3** (1979), 69–86; *MR***80b**:05025.
47. D. A. Holton, B. Manvel and B. D. McKay, Hamiltonian cycles in cubic 3-connected bipartite planar graphs, preprint.
48. D. A. Holton and B. D. McKay, The smallest non-Hamiltonian 3-connected cubic planar graphs have 38 vertices, to appear.
49. D. A. Holton, B. D. McKay, M. D. Plummer and C. Thomassen, A nine point theorem for 3-connected graphs, *Combinatorica* **2** (1982), 53–62; *MR***84i**:05069.
50. R. Honsberger, *Mathematical Gems from Elementary Combinatorics, Number Theory, and Geometry*, Dolciani Math. Expositions No. 1, Math. Assoc. of America, Buffalo, NY, 1973; *MR*54#7150.
51. M. Horňák, A theorem on nonexistence of a certain type of nearly regular cell-decompositions of the sphere, *Časopis Pěst. Mat.* **103** (1978), 333–338, 408; *MR***80h**:57007.
52. S. Jendrol', A new proof of Eberhard's theorem, *Acta Fac. Univ. Math.* **31** (1975), 1–9; *MR*52#6577.
53. S. Jendrol', On the face-vectors and vertex-vectors of maps, *Proc. Colloq. Math. Soc. János Bolyai* **18** (1976), 629–633.

54. S. Jendrol', On the face-vectors of trivalent convex polyhedra, *Math. Slovaca* **33** (1983), 165–180; *MR***85f**:52018.
55. S. Jendrol' and M. Tkáč, On the simplicial 3-polytopes with only two types of edges, *Discrete Math.* **48** (1984), 229–241; *MR***85j**:05015.
56. P. Joffe, Some properties of polytopal graphs, Ph.D. Thesis, CUNY Graduate Center, New York, 1983.
57. E. Jucovič, On the number of hexagons in a map, *J. Combinatorial Theory (B)* **10** (1971), 232–236; *MR***43**#3910.
58. E. Jucovič, On the face vector of a 4-valent 3-polytope, *Studia Sci. Math. Hungar.* **8** (1973), 53–57; *MR***48**#12304.
59. V. Klee, The use of circuit codes in analog-to-digital conversion, *Graph Theory and its Applications* (ed. B. Harris), Academic Press, New York, 1970, pp. 121–131; *MR***41**#9640.
60. J. Malkevitch, A survey of 3-valent 3-polytopes with two types of faces, *Combinatorial Structures and their Applications* (ed. R. K. Guy), Gordon and Breach, New York, 1970, pp. 255–256.
61. J. Malkevitch, Properties of Planar Graphs with Uniform Vertex and Face Structure, *Memoirs Amer. Math. Soc.* **99**, Providence, Rhode Island, 1970; *MR***41**#5240.
62. J. Malkevitch, Cycle lengths in polytopal graphs, *Theory and Applications of Graphs* (ed. Y. Alavi and D. Lick), Lecture Notes in Math. **642**, Springer, Berlin, 1978, pp. 364–370; *MR***58**#10550.
63. J. Malkevitch, Spanning trees in polytopal graphs, *Ann. New York Acad. Sci.* **319** (1979), 362–367; *MR***81h**:05062.
64. J. Malkevitch, Eberhard's theorem for convex 3-polytopes, *Convexity and Related Combinatorial Geometry* (ed. D. Kay and M. Breen), Marcel Dekker, New York, 1982, pp. 209–214.
65. H. Okamura, Every simple 3-polytope of order 32 or less is Hamiltonian, *J. Graph Theory* **6** (1982), 185–196; *MR***83e**:05080b.
66. P. J. Owens, Regular planar graphs with faces of only two types and shortness parameters, *J. Graph Theory* **8** (1984), 253–275; *MR***85j**:05017.
67. D. L. Peterson, Hamiltonian cycles in bipartite plane cubic maps, Doctoral Dissertation, Texas A&M University, 1977.
68. M. D. Plummer, On the cyclic connectivity of planar graphs, *Graph Theory and Applications* (ed. Y. Alavi *et al.*), Lecture Notes in Math. **303**, Springer, Berlin, 1972, pp. 235–242; *MR***48**#10871.
69. M. D. Plummer, On the (m^+, n^-) connectivity of 3-polytopes, *Proc. Third Southeastern Conf. on Combinatorics, Graph Theory, and Computing* (ed. F. Hoffmann *et al.*), Florida Atlantic University, Boca Raton, Florida, 1972, pp. 393–408; *MR***50**#4362.
70. M. D. Plummer and N. Robertson, Path traversability in planar graphs, *Discrete Math.* **7** (1974), 289–303; *MR***49**#142.
71. M. D. Plummer and E. L. Wilson, On cycles and connectivity in planar graphs, *Canad. Math. Bull.* **16** (1973), 283–288; *MR***48**#1987.
72. M. Rosenfeld, Polytopes of constant weight, *Israel J. Math.* **21** (1975), 24–30; *MR***52**#4132.
73. M. Rosenfeld, Are all simple 4-polytopes Hamiltonian?, *Israel J. Math* **46** (1983), 161–169; *MR***85d**:05104.
74. M. Rosenfeld and D. Barnette, Hamiltonian circuits in certain prisms, *Discrete Math.* **5** (1973), 389–394; *MR***48**#173.
75. T. Saaty and P. C. Kainen, *The Four-Color Problem. Assaults and Conquest*, McGraw-Hill, New York, 1977; *MR***58**#246.
76. E. F. Schmeichel and S. L. Hakimi, On planar graphical degree sequences, *SIAM J. Appl. Math.* **32** (1977), 598–609; *MR***57**#9583.
77. G. C. Shephard, Convex polytopes with convex nets, *Math. Proc. Cambridge Phil. Soc.* **78** (1975), 389–403; *MR***52**#11738.
78. M. Skowrońska, The pancyclicity of halin graphs and their extension contractions, *Ann. Discrete Math.* **27** (1985), 179–194.

79. E. Steinitz and H. Rademacher, *Vorlesungen über die Theorie der Polyeder*, Springer, Berlin, 1934.
80. C. Thomassen, Straight line representations of infinite planar graphs, *J. London Math. Soc. (2)* **16** (1977), 411–423; *MR***80i**:05039.
81. C. Thomassen, Planar cubic hypohamiltonian and hypotraceable graphs, Preprint Series 1978/9 No. 3, Aarhus University, Aarhus, Denmark.
82. C. Thomassen, Planarity and duality of finite and infinite graphs, *J. Combinatorial Theory (B)* **29** (1980), 244–271; *MR***81j**:05056.
83. C. Thomassen, Kuratowski's theorem, *J. Graph Theory* **5** (1981), 225–241; *MR***83d**:05039.
84. C. Thomassen, A theorem on paths in planar graphs, *J. Graph Theory* **7** (1983), 169–176; *MR***84i**:05075.
85. C. Thomassen, Deformations of plane graphs, *J. Combinatorial Theory (B)* **34** (1983), 244–257; *MR***84k**:05040.
86. C. Thomassen, Plane representations of graphs, *Progress in Graph Theory* (ed. J. A. Bondy and U. S. R. Murty), Academic Press, Toronto, Ontario, 1984, pp. 43–69; *MR***86g**:05032.
87. M. Trenkler, On the face-vector of a 5-valent convex 3-polytope, *Mat. Časopis Sloven. Akad. Vied* **25** (1975), 351–360; *MR***57**#7391.
88. M. Trenkler, Convex 4-valent polytopes with prescribed types of faces, *Comm. Math. Univ. Carolinae* **25** (1984), 171–179; *MR***85k**#52005.
89. W. T. Tutte, Bridges and Hamiltonian circuits in planar graphs, *Aequat. Math.* **15** (1977), 1–33; *MR***57**#5826.
90. W. T. Tutte, *Graph Theory*, Addison-Wesley, Reading, Mass., 1984.
91. H. Whitney, A theorem on graphs, *Ann. of Math. (2)* **32** (1931), 378–390.
92. A. Yemelichev, M. M. Kovalev and M. K. Kravtsov, *Polytopes, Graphs and Optimization*, Cambridge University Press, London, 1984.
93. J. Zaks, Pairs of Hamiltonian circuits in 5-connected planar graphs, *J. Combinatorial Theory (B)* **21** (1976), 116–131; *MR***55**#180.
94. J. Zaks, Non-Hamiltonian non-Grinbergian graphs, *Discrete Math.* **17** (1977), 317–321; *MR***57**#184.
95. J. Zaks, Non-Hamiltonian simple 3-polytopes having just two types of faces, *Discrete Math.* **29** (1980), 87–101.
96. J. Zaks, Non-Hamiltonian simple planar graphs, *Ann. Discrete Math.* **12** (1982), 255–263; *MR***86j**:05095.
97. T. Zamfirescu, On longest paths and circuits in graphs, *Math. Scand.* **38** (1976), 211–239; *MR***55**#2656.
98. T. Zamfirescu, Four graphs, to appear.

9
Hypergraphs

C. BERGE

1. Introduction

Hypergraphs are natural extensions of graphs in which edges are allowed to connect more than two vertices. They have also been called *set systems*. In this chapter, we focus our attention on results which are graphical in nature.

Formally, a **hypergraph** is a finite family H of non-empty subsets of a finite set X, whose union is X; repetitions in H are allowed. The elements of X are called **vertices**, and the elements of H are called **edges**. Note that, as a consequence of this definition and in contrast to graphs, a hypergraph is specified by its set of edges. Its vertices are thus determined, since every vertex must be in at least one edge.

We shall generally denote the vertices of a hypergraph H with small letters, and the edges by capitals:

$$X = \{x_1, x_2, \ldots, x_p\} \text{ and } H = (E_1, E_2, \ldots, E_q).$$

A hypergraph is often represented by a drawing in which points in the plane represent vertices (see Fig. 1). An edge of cardinality 1 or 2 is represented, as in a graph, by a loop or arc (or segment, in the latter case), and an edge of greater cardinality is indicated by a closed curve surrounding its elements.

GRAPH THEORY, 3
ISBN 0–12–086203–4

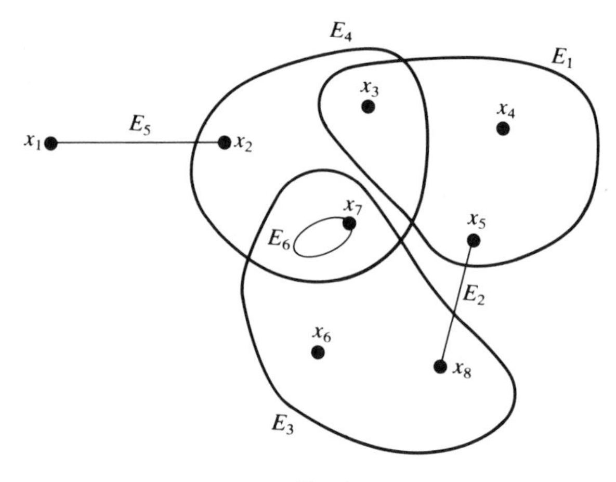

Fig. 1

In this chapter, we shall present a number of hypergraph topics, concentrating primarily on extensions of graph concepts. The next section provides the basic definitions and a few results. From there, in Sections 3 and 4, we proceed to three related concepts: edge colorings (with intersecting edges having different colors), families of edges in which every pair meet, and stars (the sets of edges containing a particular vertex). The subsequent two sections examine hypergraphs obtained by extensions (including all non-empty subsets of edges) or by truncation (chopping large edges down to subsets of fixed size). In Section 8, we resume the discussion of intersecting families, in the form of the Helly property. We then consider hypergraph counterparts of coverings and matchings, concluding with a section in which we introduce a number of further topics and present relationships among them.

2. Basic Concepts

In this section we present many of the general definitions which will be used in later sections.

If H is a hypergraph with vertex-set $X = \{x_1, x_2, \ldots, x_p\}$ and edge-set $(E_1, E_2, \ldots, E_q)$, then the **order** of H is p, and its **rank** is defined to be

$$r(H) = \max_i |E_i|.$$

A hypergraph is called **linear** if any two edges meet in at most one vertex, and **simple** (or a *Sperner family of sets*) if no edge is contained in another. A simple hypergraph is called ***r*-uniform** if all edges have cardinality r. Thus, graphs without isolated vertices are simple linear 2-uniform hypergraphs.

The counterparts of the complete graphs K_p are the **complete r-uniform hypergraphs** K_p^r; K_p^r has order p and contains all of the r-subsets of a p-set. The hypergraph generalization of the complete bipartite graph is the **complete r-partite hypergraph** $K_{p_1, p_2, \ldots, p_r}^r$, whose vertex-set X is the union of r disjoint sets $X_1, X_2, \ldots, X_r$, with $|X_i| = p_i$, and whose edge-set is the family of all r-subsets of X containing one element from each X_i.

Like graphs, hypergraphs can be represented by matrices. The **incidence matrix** $\mathbf{M}(H) = (m_{ij})$ of the hypergraph H has rows representing the vertices of H and columns representing the edges, with

$$m_{ij} = \begin{cases} 1, & \text{if } x_i \in E_j, \\ 0, & \text{otherwise.} \end{cases}$$

The **dual** of the hypergraph $H = \{E_1, E_2, \ldots, E_q\}$ with vertex-set $X = \{x_1, x_2, \ldots, x_p\}$ is the hypergraph H^* whose vertices $e_1, e_2, \ldots, e_q$ correspond to the edges $E_1, E_2, \ldots, E_q$ of H, and whose edges are the sets

$$X_i = \{e_j : x_i \in E_j \text{ in } H\}.$$

Clearly, the incidence matrix $\mathbf{M}(H^*)$ is the transpose of $\mathbf{M}(H)$, and $(H^*)^* \cong H$. Figure 2 shows the dual of the hypergraph in Fig. 1.

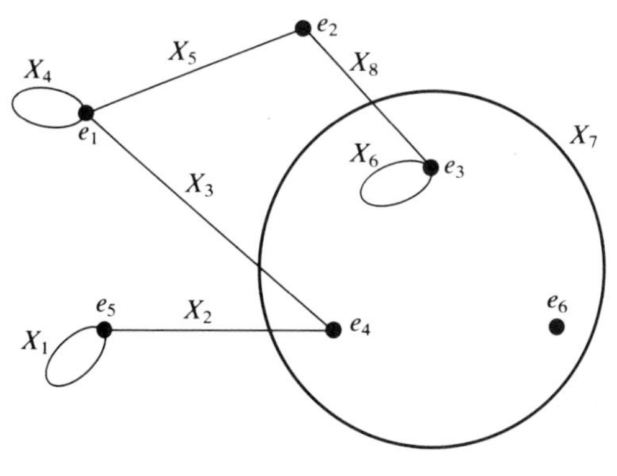

Fig. 2

The **line graph** of H, also called the *intersection graph*, is the graph $L(H)$ whose vertices correspond to the edges of H, with two vertices being adjacent if and only if the corresponding edges have a common vertex in H. Unlike the case for graphs, it is not possible to characterize these line graphs, even of 3-uniform hypergraphs, by a finite list of forbidden subgraphs.

We next consider the counterparts of a subgraph. Given a hypergraph H with vertex-set X, an **induced subhypergraph** is determined by a subset A of the vertices of H, with edge-set consisting of the family

$$H_A = (E_j \cap A \colon j \leq q \text{ and } E_j \cap A \neq \emptyset).$$

A **partial hypergraph** of H is the hypergraph generated by the edges in some subset $H' \subseteq H$, its vertices being those in some element of H'. Note that, if H is simple, then every partial hypergraph of H is also simple, but induced subhypergraphs need not be. Note also that the dual of a partial hypergraph of H is an induced subhypergraph of H^*.

A **star** of H with *center* v is any family of edges containing a given vertex v; the star consisting of all of the edges containing v is denoted by $H(v)$.

The **degree** $d(v)$ of a vertex v in a hypergraph H is the number of edges containing it. No necessary and sufficient condition is known for the degree-sequences of r-uniform hypergraphs; such a condition would be an extension of the Erdős–Gallai theorem on degree-sequences in graphs.

One can also try to characterize the lists of edge sizes $|E_1|, |E_2|, \ldots, |E_q|$ of simple hypergraphs. The following result of Sperner **[40]** gives necessary conditions for such lists; it has since been generalized in various ways:

Theorem 2.1 (Sperner's Theorem). *Let H be a simple hypergraph of order p. Then*

$$\sum_{E \in H} \binom{p}{|E|}^{-1} \leq 1, \tag{1}$$

and the number of edges in H satisfies

$$q(H) \leq \binom{p}{\lfloor \frac{1}{2}p \rfloor}. \tag{2}$$

Furthermore, equality holds in (2) *if and only if H is the complete h-uniform hypergraph K_p^h, with $h = \frac{1}{2}p$ if p is even, and $h = \frac{1}{2}(p-1)$ or $h = \frac{1}{2}(p+1)$ if p is odd.* ||

The following corollary is an immediate consequence of inequality (2):

Corollary 2.2. *Every simple hypergraph H of order p and rank $r \leq \frac{1}{2}p$ has at most $\binom{p}{r}$ edges, and has exactly $\binom{p}{r}$ edges if and only if $H \cong K_p^r$.* ||

A hypergraph H is said to be a **separating hypergraph** if the intersection of all edges containing any given vertex v is $\{v\}$. For example, a graph is a separating hypergraph if every vertex has degree at least 2.

Corollary 2.3. *The degrees $d_1, d_2, \ldots, d_p$ of a separating hypergraph with q edges and p vertices satisfy the inequality*

$$\sum_{i=1}^{p} \binom{q}{d_i}^{-1} \leq 1.$$

Proof. Since H is a separating hypergraph if and only if its dual H^* is simple, the result follows from inequality (1) of Sperner's theorem. ||

3. Intersecting Families

An **intersecting family** in a hypergraph is the partial hypergraph of a collection of pairwise intersecting edges. In a graph, the only intersecting families are the stars and triangles. We let $\Delta_0(H)$ denote the maximum number of edges in an intersecting family of H. Clearly, $\Delta_0(H) \geq \Delta(H)$, the maximum degree of H. We say that H has the **star-intersection property** if equality holds.

Theorem 3.1. *If H is a hypergraph of order p with no repeated edges, then $\Delta_0(H) \leq 2^{p-1}$.*

Proof. Let X be a set of order p, and let $\mathcal{A}$ be a maximal intersecting family for the hypergraph of all non-empty subsets of X. Let B be a subset of X not in $\mathcal{A}$. Then by the maximality of $\mathcal{A}$ there exists a set $A \in \mathcal{A}$ with $B \cap A = \emptyset$. Hence, $A \subseteq X - B$, and so $(X - B) \cap S \neq \emptyset$ for all $S \in \mathcal{A}$. Again, by the maximality of $\mathcal{A}$, this means that $X - B \in \mathcal{A}$. Conversely, if $X - B \in \mathcal{A}$, then $B \notin \mathcal{A}$, and so $B \leftrightarrow X - B$ is a one-to-one correspondence between $\mathcal{P}(X) - \mathcal{A}$ and $\mathcal{A}$. Consequently, $|\mathcal{A}| = \frac{1}{2}|\mathcal{P}(X)| = 2^{p-1}$. ||

If a hypergraph itself is an intersecting family, we call it an **intersecting hypergraph**. The following theorem was first discovered by Erdős, Ko and Rado [**16**] in 1961. It has undergone numerous improvements and generalizations—see, for example, Hilton and Milner [**26**], Schönheim [**37**], Bollobás [**9**], Frankl [**19**], [**20**], Greene *et al.* [**22**], Hilton [**25**], and Erdős *et al.* [**17**]:

Theorem 3.2. *Let H be a simple intersecting hypergraph of order p and rank $r \leq \frac{1}{2}p$. Then*

$$\sum_{E \in H} \binom{p-1}{|E|-1}^{-1} \leq 1.$$

Furthermore, the number of edges in H satisfies

$$q(H) \leq \binom{p-1}{r-1},$$

with equality holding when H is a star of K_p^r, and only then if $r < \frac{1}{2}p$. ||

The following corollary considers the intersecting family number of the complete r-uniform hypergraphs:

Corollary 3.3.

$$\Delta_0(K_p^r) = \begin{cases} \binom{p-1}{r-1} = \Delta(K_r^p), & \text{for } r \leq \frac{1}{2}p, \\ \binom{p}{r}, & \text{for } r > \frac{1}{2}p. \end{cases}$$

Furthermore, the only maximum intersecting families of K_p^r are

stars, if $r < \frac{1}{2}p$;

the maximal intersecting families, if $r = \frac{1}{2}p$; and

itself, if $r > \frac{1}{2}p$. ||

4. The Edge-coloring Property

The **chromatic index** $\chi'(H)$ of a hypergraph H is the least number of colors needed to color the edges of H so that no two intersecting edges have the same color. Clearly,

$$\chi'(H) \geq \Delta_0(H) \geq \Delta(H).$$

If equality holds throughout, then H is said to have the **edge-coloring property**. Clearly, this is stronger than the star-intersection property $\Delta_0(H) = \Delta(H)$ mentioned in the last section.

Graphs with the edge-coloring property are often called **class 1** graphs. By König's theorem, all bipartite graphs have this property. A result of Gallai shows that the same is true of all interval hypergraphs, these being the hypergraphs whose vertices are points on a line and whose edges are determined by intervals on that line.

Theorem 4.1. *Every interval hypergraph has the edge-coloring property.* ||

It is not difficult to show that complete graphs of even order also have the property (it was known to Lucas), and Peltesohn [**36**] showed that the complete 3-uniform hypergraphs K_{3p}^3 do also. These results were extended to the general case by Baranyai [**1**]:

Theorem 4.2. *The complete r-uniform hypergraph K_p^r has the edge-coloring property if and only if p is divisible by r.* ||

We shall prove the corresponding result for complete r-partite hypergraphs:

Theorem 4.3. *For $r \geq 2$, the complete r-partite hypergraph $K^r_{p_1, p_2, \ldots, p_r}$ has the edge-coloring property.*

Proof. Assume that $p_1 \leq p_2 \leq \ldots \leq p_r$, and let the elements of the ith partite set X_i be $0, 1, 2, \ldots, p_i - 1$. As usual, let $[n]_k$ denote the residue class congruent to n (modulo k), where n lies between 0 and $k - 1$. To each edge $\bar{x} = \{x_1, x_2, \ldots, x_r\}$, where $x_i \in X_i$, assign the $(r - 1)$-tuple

$$\zeta(\bar{x}) = ([x_1 + x_2]_{p_2}, [x_1 + x_3]_{p_3}, \ldots, [x_1 + x_r]_{p_r}).$$

Assume now that $\bar{y} = \{y_1, y_2, \ldots, y_r\}$ and $\bar{z} = \{z_1, z_2, \ldots, z_r\}$ are distinct edges with non-empty intersection. If $y_1 = z_1$, then there exists an integer $i > 1$, with $y_i \neq z_i$, whence $[y_1 + y_i]_{p_i} \neq [z_1 + z_i]_{p_i}$ and $\zeta(\bar{y}) \neq \zeta(\bar{z})$. On the other hand, if $y_1 \neq z_1$, then there exists $i > 1$ with $y_i = z_i$, so $\zeta(\bar{y}) \neq \zeta(\bar{z})$ just as before. Therefore, the mapping $\bar{x} \rightarrow \zeta(\bar{x})$ assigns to each edge a 'color' in such a way that intersecting edges have different colors. Therefore, the chromatic index is at most $p_2 p_3 \ldots p_r$, which is $\Delta(K^r_{p_1, p_2, \ldots, p_r})$. ||

5. Hereditary Closures

In this section we consider the above two properties for hypergraphs with lots of edges. The **hereditary closure** $\hat{H}$ of a hypergraph H consists of all non-empty subsets (without repetition) of edges of H.

Chvátal **[14]** has conjectured that the hereditary closures of all hypergraphs satisfy the star-intersection property—that is, any $\hat{H}$ always has at least one star among the maximum intersecting families of any hypergraph H:

Conjecture 5.1 (Chvátal's Conjecture). *Every hypergraph H satisfies $\Delta_0(\hat{H}) = \Delta(\hat{H})$.*

We also conjecture that the hereditary closure of every linear hypergraph (one with no two edges meeting in more than one vertex) satisfies the edge-coloring property:

Conjecture 5.2. *Every linear hypergraph H satisfies $\chi'(\hat{H}) = \Delta(\hat{H})$.*

It follows from Vizing's theorem on the chromatic index of graphs (which are linear hypergraphs) that their hereditary closures always have the edge-coloring property:

Theorem 5.1. *The hereditary closure of every graph has the edge-coloring property.*

Proof. For any graph G, $\Delta(\hat{G}) = \Delta(G) + 1$, and by Vizing's theorem, $\chi'(G) = \Delta(G)$ or $\Delta(G) + 1$. In either case, consider any optimal edge-coloring of G. If $\Delta(G) + 1$ colors are used in coloring the edges of $\hat{G}$, then there is always one available for each of the loops. Hence, $\chi'(G) \leq \Delta(G) + 1$, and the result follows. ||

Thus, if Conjecture 5.2 is true, then it is a generalization of Vizing's theorem.

The next theorem, which was proved first for uniform hypergraphs by Sterboul [**44**] and later in the general case by Stein [**41**], shows that the overlapping portion of the two conjectures is true:

Theorem 5.2. *The hereditary closure of every linear hypergraph satisfies the star-intersection property.* ||

Other families of hypergraphs for which the same conclusion holds are given in the next two results. The first was found independently by Stein and Schönheim [**42**] and by Wang and Wang [**46**], while the second was the discovery of Sterboul [**44**]:

Theorem 5.3. *The hereditary closure of every hypergraph of maximum degree* 2 *satisfies the star-intersection property.* ||

Theorem 5.4. *The hereditary closure of every* 3*-uniform hypergraph satisfies the star-intersection property.* ||

This theorem cannot be extended to the edge-coloring property; for example, the hypergraph $\hat{K}_7^3$ has chromatic index greater than its maximum degree. Baranyai's theorem (Theorem 4.2) shows that the hereditary closure of the general complete r-uniform hypergraph K_p^r has the edge-coloring property if and only if, from a stock of rods of length p, one can cut without waste $\binom{p}{\lambda}$ sticks of length λ, for $\lambda = 1, 2, \ldots, p$—that is, the *cutting stock problem* must be solvable. For rank 4, the hypergraphs $\hat{K}_9^4$ and $\hat{K}_{10}^4$ do not have the edge-coloring property, but as a consequence of Theorem 3.2 (the theorem of Erdős–Ko–Rado), $\hat{K}_p^r$ has the star-intersection property.

The case of complete r-partite hypergraphs was settled by Berge and Johnson [**6**]:

Theorem 5.5. *The hereditary closure of every complete r-partite hypergraph has the edge-coloring property.* ||

In 1973, Schönheim [**37**] proved that the hereditary closure of every star has the star-intersection property, a result that has been extended to the edge-coloring property by Berge [**3**]:

Theorem 5.6. *The hereditary closure of every star has the edge-coloring property.*

Proof. Let H be a star. By a result of Berge [**2**], the complement of the intersection graph of the family $\hat{H} \cup \{\varnothing\}$ has a perfect matching, and this yields an edge-coloring of H with $\frac{1}{2}(|\hat{H}| + 1)$ colors, using each color twice. Furthermore, for any center v of the star H, all of the edges of H which contain v have different colors. Since the mapping $E \leftrightarrow E - \{v\}$ is a one-to-one correspondence between $\hat{H}(v)$ and $(\hat{H} \cap \{\varnothing\}) - \hat{H}(v)$, it follows that $|\hat{H}(v)| = \frac{1}{2}(|\hat{H}| + 1)$. Therefore, $\chi'(\hat{H}) \leq |\hat{H}(v)| \leq \Delta(\hat{H})$, which completes the proof. ||

6. *k*-sections

Let H be a simple hypergraph of order p and rank r on the set X. For $k \leq r$, the ***k*-section** of H, denoted by $[H]_k$, is the hypergraph on X whose edge-set is

$$\{F \in H: |F| < k\} \cup \{F: |F| = k \text{ and } F \subseteq E \text{ for some } E \in H\}.$$

Clearly, the r-section of H is H itself and, for $k = r - 1$, the number of edges satisfies $q[H]_{r-1} \leq rq(H)$. On the opposite side, the best possible lower bound was obtained independently by Kruskal [**29**] and Katona [**28**]. For its statement, we define a finite sequence of integers $(a_1, a_2, \ldots, a_s)$, using the rank r and the number q of edges as follows:

a_r is the greatest integer for which $\binom{a_r}{r} \leq q$;

if $a_r, \ldots, a_{r-k}$ have been defined and $q = \binom{a_r}{r} + \ldots + \binom{a_{r-k}}{r-k}$,

let $s = r - k$;

otherwise, let a_{r-k-1} be the greatest integer for which

$$\binom{a_r}{r} + \ldots + \binom{a_{r-k}}{r-k} + \binom{a_{r-k-1}}{r-k-1} \leq q.$$

Theorem 6.1. *Let H be an r-uniform hypergraph with q edges, and let $a_r \geq \ldots \geq a_s \geq s \geq 1$ be the sequence for which*

$$q(H) = \binom{a_r}{r} + \binom{a_{r-1}}{r-1} + \ldots + \binom{a_s}{s}.$$

Then

$$q([H]_{r-1}) \geq \binom{a_r}{r-1} + \binom{a_{r-1}}{r-2} + \ldots + \binom{a_s}{s-1}. \ \|$$

Corollary 6.2. *With the notation of Theorem* 6.1,

$$q([H]_k) \geq \binom{a_r}{k}. \ \|$$

The 2-section of a hypergraph H is a graph, with no loops if H has no singleton edges. H is called **conformal** if every maximal clique of $[H]_2$ defines an edge of H. (Equivalently, a simple hypergraph is conformal if and only if its edges are the maximal cliques of some graph.) More generally, H is said to be ***k*-conformal** if $[H]_k$ has the property that each maximal complete subhypergraph defines an edge of H. The following characterization of k-conformal hypergraphs is due to Berge and Duchet [5] (generalizing an earlier result of Gilmore):

Theorem 6.3. *A simple hypergraph H is k-conformal if and only if, for each partial hypergraph H' with $k + 1$ edges, the set $\{v: d_{H'}(v) \geq k\}$ is contained in some edge of H.* $\|$

7. The Helly Property

In this section we consider another intersection property: a hypergraph has the **Helly property** if every intersecting family is a star. For a graph, the Helly property holds if and only if the graph is triangle-free.

Other examples of hypergraphs with the Helly property are the interval hypergraphs of points on a line. More generally, for any lattice $(X, \leq)$, any family of intervals $[a, b] = \{x: a \leq x \leq b\}$ has the Helly property. Furthermore, for any tree T, the sets of vertices of its subtrees form a hypergraph with the Helly property.

For any integer $k \geq 2$, a hypergraph H is said to have the ***k*-Helly property** if, for every set J of edges, the following two conditions are equivalent:

(*i*) if $I \subseteq J$ and $|I| \leq k$, then $\bigcap_{i \in I} E_i \neq \emptyset$;

(*ii*) $\bigcap_{i \in J} E_i \neq \emptyset$.

It is readily seen that the usual Helly property is the case $k = 2$.

It can also be seen that a hypergraph H has the k-Helly property if and only if its dual H^* is k-conformal. The following result is therefore a consequence of Theorem 6.3:

Theorem 7.1. *A hypergraph H has the k-Helly property if and only if, for every set A of $k + 1$ vertices, the intersection of those edges E such that $|E \cap A| \geq k$ is non-empty.* ||

For the Helly property itself, this is simply the statement that, for any three vertices, those edges containing at least two of them have non-empty intersection.

The following corollary, giving an upper bound on the number of edges in certain hypergraphs with the k-Helly property, was first found by Bollobás and Duchet [**10**]:

Corollary 7.2. *Let H be a hypergraph of order p and rank r, which has the k-Helly property. If every edge has at least $k + 2$ vertices, then the number of edges in H is at most $\binom{p-1}{r-1}$.* ||

Corollary 7.3. *Let H be a simple hypergraph of order p and rank r, with $3 \leq r \leq \frac{1}{2}p$, and having the Helly property. Then H has at most $\binom{p-1}{r-1}$ edges, with equality holding if and only if H is a star of K_p^r.* ||

Our last result in this section can be used to provide characterizations of the line graphs (intersection graphs) of certain families of hypergraphs:

Theorem 7.4. *The line graph of a hypergraph H is isomorphic to the 2-section of the dual H^*—that is, $L(H) \cong [H^*]_2$.* ||

8. Transversal Numbers

A **transversal set** of a hypergraph H is a set of vertices which meets all the edges of H; this concept is thus an extension of the vertex-cover of a graph. The least order of such a set is denoted by $\tau(H)$, and the greatest order of a transversal set which is minimal (with respect to inclusion) is denoted by $\tau'(H)$.

These parameters have been extensively studied for many structures—for example, projective planes (see Pelikán [**35**]), affine planes (see Brouwer and Schrijver [**12**]) and block designs. Many bounds, both upper and lower, are known; they are usually obtained by using either probabilistic methods or classical results of 'fractional graph theory' (see Berge [**4**]).

The following result gives an upper bound for τ':

Theorem 8.1. *If H is a hypergraph of order p, minimum edge-size $s > 1$, and maximum degree Δ, then*

$$\tau'(H) \leq \frac{p\Delta}{\Delta + s - 1}. \quad \|$$

This result is best possible for $s = 2$, but not for $s = 3$. Stronger results have been found by Meyer [**34**].

If H is a hypergraph on X in which X is not an edge, then the **edge-complement** $\bar{H}$ is a hypergraph on X whose edges are the sets $X - E$, for E in H. The following result is due to Berge and Duchet [**5**]:

Theorem 8.2. *Let H be a hypergraph in which no edge has all the vertices. Then $\tau'(H) \leq k$ if and only if $\bar{H}$ is k-conformal.* ||

When combined with Theorem 6.3, this result gives a method for determining whether $\tau'(H) \leq k$ for some classes of hypergraphs.

The following upper bound for τ was found by both Stein [**41**] and Lovász [**30**]:

Theorem 8.3. *For any hypergraph H,*

$$\tau(H) \leq \left(1 + \tfrac{1}{2} + \ldots + \frac{1}{\Delta(H)}\right) \max_{H' \subseteq H} \frac{q(H')}{\Delta(H')}. \quad \|$$

Corollary 8.4. *If H is a Δ-regular r-uniform hypergraph of order p, then*

$$\tau(H) \leq (1 + \log \Delta)p/r. \quad \|$$

A hypergraph H is said to be **τ-critical** if, for every edge E of H, $\tau(H - E) < \tau(H)$. Every hypergraph H with $\tau(H) = t$ has a partial hypergraph H' with $\tau(H') = t$ which is τ-critical. Examples of τ-critical hypergraphs include edge-critical graphs with no isolated vertices, and complete hypergraphs. The following theorem was found by Tuza [**45**]:

Theorem 8.5. *If H is a τ-critical hypergraph with $\tau(H) = t + 1$, then*

$$\sum_{E \in H} \binom{|E| + t}{t}^{-1} \leq 1. \quad \|$$

An upper bound on the number of edges in a τ-critical hypergraph can be deduced as a corollary; it was found by Bollobás [**8**] and Jaeger and Payan [**27**]:

Corollary 8.6. *A τ-critical hypergraph of rank r with $\tau(H) = t + 1$ has at most $\binom{r + t}{t}$ edges.* ||

Clearly, the bound in this corollary is attained by K_{r+t}^r. Other properties, many similar to properties of edge-critical graphs, have been discussed by Gyárfás *et al.* [**23**].

We conclude this section with a related result on coverings. The **vertex-covering number** $\alpha(H)$ is the minimum order of a set of edges which contain all the vertices of H among them; since $\alpha(H) = \tau(H^*)$, the next statement follows from Theorem 8.2:

Theorem 8.9. *In a hypergraph of rank r with q edges and minimum degree* $\delta \geqslant 2$, *the vertex-covering number is at most* $qr/(r + \delta + 1)$. ||

9. The König Property

As in a graph, edges in a hypergraph H are said to be **independent** if they have no vertices in common. A **matching** of H is then a partial hypergraph whose edges are independent, and the **matching number** $\nu(H)$ (or *edge-independence number*) is the maximum cardinality of a matching. Clearly, $\tau(H) \geq \nu(H)$.

If $\nu(H) = \tau(H)$, then H is said to have the **König property**. Note that complete r-uniform hypergraphs do not have this property for $r > 1$, since $\nu(K_p^r) = \lfloor p/r \rfloor$ and $\tau(K_p^r) = p - r + 1$. However, all complete r-partite hypergraphs have the property, since $\nu = \tau = \min p_i$ in $K_{p_1, p_2, \ldots, p_r}^r$. In this area, H. Ryser has made the following conjecture:

Conjecture 9.1. *If H is a partial hypergraph of some complete r-partite hypergraph, then* $\tau(H) \leq (r - 1)\nu(H)$.

Consider now the following packing problem for rectangular bricks:

> *given a* $p \times q$ *chessboard of unit squares and a supply of bricks of size* $a \times b$, *what is the largest number of bricks that can be placed on the chessboard?*

The answer is given by the matching number of an appropriate hypergraph of order pq. It was shown by Brualdi and Foregger [**13**] that this hypergraph has the König property for all p and q if and only if a is a divisor of b, or vice versa.

Many bounds on the matching number have been obtained, usually using extremal or fractional graph theory. We mention just one of these results here, due to Seymour [**39**]:

Theorem 9.2. *If a linear hypergraph H with p vertices and q edges has no multiple loops, then its matching number is at least* q/p. ||

Those hypergraphs for which equality holds in Theorem 9.2 were also determined by Seymour.

10. A Variety of Hypergraph Properties

In concluding this chapter, we take a look at a number of other families of hypergraphs; our main objective is to present the relationships among them. We shall also connect them with a couple of the properties encountered in earlier sections, and in some cases, characterizations will be given.

We begin with a couple of basic definitions. The **vertex-chromatic number** $\chi(H)$ of a hypergraph H is the smallest number of colors needed to color the vertices so that no non-trivial edge has all of its vertices assigned the same color. (A *strong chromatic number*, for which all vertices in each edge must have different colors, has also been studied, but we shall not have use for it here.) A **cycle** is a sequence $(v_1, E_1, v_2, E_2, \ldots, v_k, E_k, v_1)$, with $k \geq 3$, in which $v_1, v_2, \ldots, v_k$ are distinct vertices, $E_1, E_2, \ldots, E_k$ are distinct edges and, for each i, E_i contains v_i and v_{i+1}.

We now present nine more families of hypergraphs:

(A) **Cycle-free hypergraphs.** These are generalizations of forests.

(B) **Even-cycled hypergraphs.** These generalize bipartite graphs.

(C) **Unimodular hypergraphs.** Such a hypergraph has a unimodular incidence matrix—that is, each of the subdeterminants is 0, 1 or -1. Two examples of these structures are found in a tree T in which all edges have been arbitrarily oriented. One unimodular hypergraph is on the vertex-set of T and has edges given by the directed paths of the tree. Another is on the set of directed edges of T and has edges given by the directed paths of T. That these hypergraphs are unimodular follows from the following theorem of Ghouila-Houri [**21**]:

Theorem 10.1. *A hypergraph $H = (E_1, E_2, \ldots, E_q)$ on X is unimodular if and only if, for every $A \subseteq X$, the subhypergraph H_A induced by A has a 2-coloring which is 'equitable', in the sense that the numbers of vertices of the two colors in each edge are as nearly equal as possible.* ||

(D) **Balanced hypergraphs.** These are hypergraphs in which each odd cycle has an edge containing at least three vertices of the cycle. An example is shown in Fig. 3. Most of the usual minimax properties hold for balanced hypergraphs. In the following theorem, the equivalence of properties (i) and (iii) is due to Berge and Las Vergnas [**7**], and that of properties (i) and (iv) is due to Berge [**4**]:

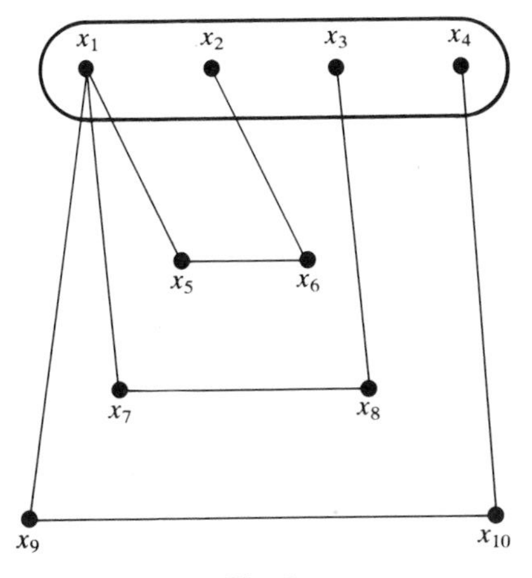

Fig. 3

Theorem 10.2. *The following statements are equivalent for any hypergraph H:*

(*i*) *H is balanced;*

(*ii*) *every induced subhypergraph is vertex 2-colorable;*

(*iii*) *every partial subhypergraph has the König property;*

(*iv*) *every partial subhypergraph has the edge-coloring property.* ||

(*E*) **Normal hypergraphs**. A hypergraph is **normal** if every partial hypergraph has the König property. The following characterizations of normal hypergraphs are part of Lovász' famous theorem **[30]** on perfect graphs:

Theorem 10.3. *The following statements are equivalent for any hypergraph H:*

(*i*) *H is normal;*

(*ii*) *every partial hypergraph of H has the edge-coloring property;*

(*iii*) *H is the dual of the family of maximal cliques of a perfect graph.* ||

(*F*) **Pseudo-balanced hypergraphs**. These are hypergraphs in which every odd cycle has three edges with a non-empty intersection. A characterization of these structures has been given by Fournier and Las Vergnas **[18]**.

(*G*) **Totally balanced hypergraphs**. These are defined by every cycle having an edge containing at least three vertices of the cycle.

(*H*) **Tree-like hypergraphs**. A hypergraph is **tree-like** if it has the Helly property and is pseudo-balanced. The reason for the name is that *a*

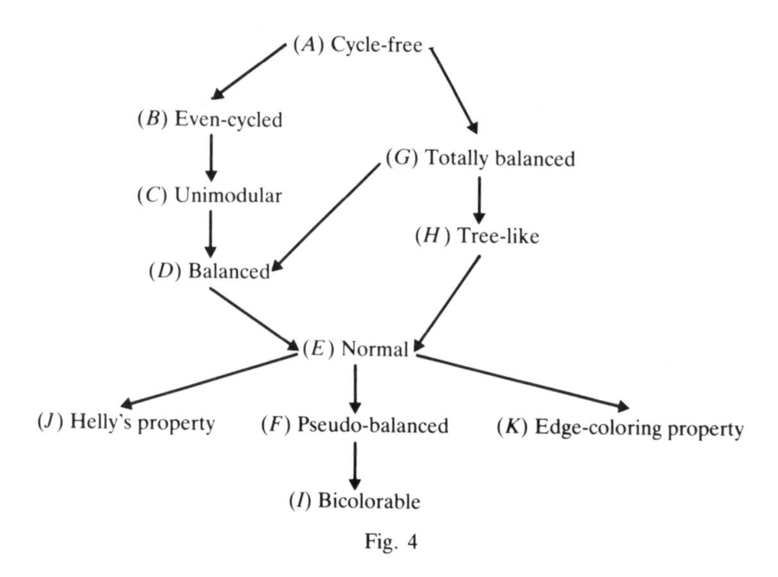

Fig. 4

hypergraph is tree-like if and only if it is the family of subtrees of a tree. These hypergraphs are also precisely the duals of the family of cliques of a triangulated graph.

(*I*) **Bicolorable hypergraphs**. These are hypergraphs with vertex-chromatic number $\chi \leqslant 2$.

The diagram in Fig. 4 shows the relationships between these nine families and the families with the edge-coloring property and Helly's property. Needless to say, more could be added, but these give a representative idea of the fruitful ideas inherent in set systems, or hypergraphs.

References

1. Z. Baranyai, On the factorization of the complete uniform hypergraph, *Infinite and Finite Sets I* (ed. A. Hajnal, R. Rado and V. T. Sós), North-Holland, Amsterdam, 1975, pp. 91–108; *MR***54**#5047.
2. C. Berge, The rank of a family of sets and some applications to graph theory, *Recent Progress in Combinatorics* (ed. W. T. Tutte), Academic Press, New York, 1969, pp. 49–57; *MR***41**#5231.
3. C. Berge, A theorem related to the Chvátal conjecture, *Proc. Fifth British Combinatorial Conference* (ed. C. St. J. A. Nash-Williams and J. Sheehan), *Congressus Numerantium XV*, Utilitas Math., Winnipeg, 1976, pp. 35–40; *MR***53**#13035.
4. C. Berge, Packing problems and hypergraph theory, *Ann. Discrete Math.* **4** (1979), 3–37.
5. C. Berge and P. Duchet, A generalization of Gilmore's theorem, *Recent Advances in Graph Theory* (ed. M. Fiedler), Academia, Prague, 1975, pp. 49–55; *MR***53**#10585.
6. C. Berge and E. L. Johnson, Coloring the edges of a hypergraph and linear programming techniques, *Studies in Integer Programming*, *Ann. Discrete Math.* **1** (1977), 65–78; *MR***58**#16406.

7. C. Berge and M. Las Vergnas, Sur un théorème du type König pour hypergraphes, *Int. Conf. on Combinatorial Math.* (ed. A. Gewirtz and L. Quintas), *Ann. New York Acad. Sci.* **175** (1970), 32–40; *MR***42**#1690.
8. B. Bollobás, On generalized graphs, *Acta Math. Acad. Sci. Hungar.* **16** (1965), 447–452; *MR***32**#1133.
9. B. Bollobás, Sperner systems consisting of pairs of complementary subsets, *J. Combinatorial Theory (A)* **15** (1973), 363–366; *MR***48**#5867.
10. B. Bollobás and P. Duchet, Helly families of maximal size, *J. Combinatorial Theory (A)* **26** (1979), 197–200; *MR***80j**:05004.
11. B. Bollobás and P. Duchet, On Helly families of maximal size, *J. Combinatorial Theory (B)* **35** (1983), 290–296; *MR***85d**:05004.
12. A. E. Brouwer and A. Schrijver, The blocking number of an affine space, *J. Combinatorial Theory (A)* **24** (1978), 251–253; *MR***58**#293.
13. R. A. Brualdi and T. H. Foregger, Packing boxes with harmonic bricks, *J. Combinatorial Theory (B)* **17** (1974), 81–114.
14. V. Chvátal, Intersecting families of edges in hypergraphs having the hereditary property, *Hypergraph Seminar* (ed. C. Berge and D. K. Ray-Chaudhuri), Lecture Notes in Math. **411**, Springer, Berlin, 1974, pp. 61–66; *MR***51**#1405.
15. D. E. Daykin, A. J. W. Hilton and D. Miklós, Pairings from down-sets and up-sets in distributive lattices, *J. Combinatorial Theory (A)* **35** (1983), 215–230; *MR***85f**:06016.
16. P. Erdős, Chao-Ko and R. Rado, Intersection theorems for systems of finite sets, *Quart. J. Math. (Oxford) (2)* **12** (1961), 313–320; *MR***25**#3839.
17. P. L. Erdős, P. Frankl and G. O. H. Katona, Intersecting families and their convex hull, *Combinatorica* **4** (1984), 21–34; *MR***85j**:05026.
18. J.-C. Fournier and M. Las Vergnas, Une classe d'hypergraphes bichromatiques, *Discrete Math.* **2** (1972), 407–410; *MR***46**#5156.
19. P. Frankl, On intersecting families of finite sets, *J. Combinatorial Theory (A)* **24** (1978), 146–161; *MR***58**#250.
20. P. Frankl, Families of finite sets satisfying a union condition, *Discrete Math.* **21** (1979), 111–118.
21. A. Ghouila-Houri, Charactérisation des matrices totalement unimodulaires, *C. R. Acad. Sci. Paris (A)* **254** (1962), 1192–1194; *MR***24**#A2589.
22. C. Greene, G. O. H. Katona and D. J. Kleitman, Extensions of the Erdős–Ko–Rado theorem, *Recent Advances in Graph Theory* (ed. M. Fiedler), Academia, Prague, 1975, pp. 223–231, and *SIAM* **55** (1976), 1–8; *MR***56**#15438.
23. A. Gyárfás, J. Lehel and Z. Tuza, Upper bound on the order of τ-critical hypergraphs, *J. Combinatorial Theory (B)* **33** (1982), 161–165.
24. P. Hansen and M. Loréa, Degrees and independent sets of hypergraphs, *Discrete Math.* **14** (1976), 305–309; *MR***52**#10505.
25. A. J. W. Hilton, An intersection theorem for a collection of families of subsets of a finite set, *J. London Math. Soc. (2)* **15** (1977), 369–376; *MR***56**#2834.
26. A. J. W. Hilton and E. C. Milner, Some intersection theorems for systems of finite sets, *Quart. J. Math. (Oxford) (2)* **18** (1967), 369–384; *MR***36**#2510.
27. F. Jaeger and C. Payan, Détermination du nombre maximum d'arêtes d'un hypergraphe τ-critique, *C. R. Acad. Sci. Paris (A)* **271** (1971), 221–223.
28. G. O. H. Katona, A theorem of finite sets, *Theory of Graphs* (ed. P. Erdős and G. Katona), Akadémia Kiadó, Budapest, 1968, pp. 187–207.
29. J. B. Kruskal Jr., The number of simplices in a complex, *Mathematical Optimization Techniques*, University of California Press, Berkeley, 1963, pp. 251–278; *MR***27**#4771.
30. L. Lovász, Normal hypergraphs and the perfect graph conjecture, *Discrete Math.* **2** (1972), 253–267; *MR***46**#1624.
31. L. Lovász, On the ratio of optimal integral and fractional covers, *Discrete Math.* **13** (1975), 383–390; *MR***52**#5452.
32. L. Lovász, *Combinatorial Problems and Exercises*, North-Holland, Amsterdam, 1979; *MR***80m**:05001.
33. D. Lubell, A short proof of Sperner's theorem, *J. Combinatorial Theory* **1** (1966), 299; *MR***33**#2558.

34. J.-C. Meyer, Ensembles stables maximaux dans les hypergraphes, *C. R. Acad. Sci. Paris (A)* **274** (1975), 144–147; *MR***44**#6508.
35. J. Pelikán, Properties of balanced incomplete block designs, *Combinatorial Theory and its Applications* (ed. P. Erdős, A Rényi and V. T. Sós), North-Holland, Amsterdam, 1971, pp. 869–890.
36. R. Peltesohn, Das Turnierproblem, Thesis, Friederich Wilhelms Universität, Berlin, 1983.
37. J. Schönheim, On a problem of Daykin concerning intersecting families of sets, *Combinatorics* (ed. T. McDonough and V. C. Mavron), London Math. Soc. Lecture Notes **13**, Cambridge University Press, 1974, pp. 139–140; *MR***50**#9598.
38. J. Schönheim, Hereditary systems and Chvátal's conjecture, *Proc. Fifth British Combinatorial Conference* (ed. C. St. J. A. Nash-Williams and J. Sheehan), *Congressus Numerantium XV*, Utilitas Math., Winnipeg, 1976, pp. 537–539; *MR***52**#13403.
39. P. D. Seymour, Packing nearly disjoint sets, *Combinatorica* **2** (1982), 91–97; *MR***83m**:05044.
40. E. Sperner, Ein Satz über Untermengen einer endlichen Menge, *Math. Z.* **27** (1928), 544–548.
41. P. Stein, Chvátal's conjecture and point-intersections, *Discrete Math.* **43** (1983), 321–323; *MR***84g**:05111.
42. P. Stein and J. Schönheim, On Chvátal's conjecture related to hereditary systems, *Ars Combinatoria* **5** (1978), 275–291.
43. S. K. Stein, Two combinatorial covering theorems, *J. Combinatorial Theory (A)* **16** (1974), 391–397.
44. F. Sterboul, Sur une conjecture de V. Chvátal, *Hypergraph Seminar* (ed. C. Berge and D. K. Ray-Chaudhuri), Lecture Notes in Math. **411**, Springer, Berlin, 1974, pp. 152–164; *MR***52**#122.
45. Z. Tuza, Helly-type hypergraphs and Sperner families, *Europ. J. Combinatorics* **5** (1984), 185–187; *MR***85k**:05081.
46. D. L. Wang and P. Wang, Some results about the Chvátal conjecture, *Discrete Math.* **24** (1978), 95–101; *MR***80b**:05003.

Index of Definitions